The Economics of the Modern Construction Firm

Stephen L. Gruneberg
Lecturer
Bartlett School
University College London

and

Graham J. Ive
Senior Lecturer
Bartlett School
University College London

D1334028

First published 2000 by
MACMILLAN PRESS LTD
Houndmills, Basingstoke, Hampshire RG21 6XS
and London
Companies and representatives
throughout the world

ISBN 0-333-79027-8 hardcover
ISBN 0-333-91995-5 paperback

A catalogue record for this book is available from the British Library.

This book is printed on paper suitable for recycling and made from fully managed and sustained forest sources.

10 9 8 7 6 5 4 3 2 1
09 08 07 06 05 04 03 02 01 00

Printed and bound in Great Britain by
Antony Rowe Ltd, Chippenham, Wiltshire

To Jan and Rose

Contents

Part II   Construction Markets

Part V Capital Circuits

List of Figures

List of Tables

Preface

This book, *The Economics of the Modern Construction Firm*, is the complement of *The Economics of the Modern Construction Sector*. Like the latter, this volume sets the context for an understanding of the operation of firms within the construction production process, but takes the analysis further into the particular decision making and planning processes of firms involved in construction.

These processes are necessarily distinct from those found in many other sectors of the economy. They are peculiar to construction primarily because of the project nature of the building and civil engineering industries. There are a variety of features of the production of the built environment which distinguish it from the norm found in most economic texts. To begin with there is the one-off nature of projects. Each project is unique in a number of ways in terms of size, contractual arrangements, design, mix of materials and technology used, methods of construction, participants, site and location.

The firms themselves are also exceptional to the paradigm assumed in much of the economics literature. Construction firms have several distinct characteristics which directly affect their *modus operandi*. Where other firms have stocks, none exist in contract construction. Where other firms have fixed costs, construction firms have variable costs. Where other firms have assets, construction firms hire. Where other firms own or rent their premises or place of production, contract construction firms do not have to pay for the site of their production. Where other firms have continuous production, or provide repetitive services, construction firms have to deal with one-off projects.

In *The Economics of the Modern Construction Firm*, we attempt to provide a realistic theoretical economic framework for understanding the firm in the construction industry. This book provides the necessary theoretical economic background for an understanding of firms operating in construction and together with *The Economics of the Modern Construction Sector*, the two books are aimed at enabling firms to improve their decision making, strategic thinking and planning effort.

This volume is divided into five parts entitled

- social structures of accumulation,
- construction markets,

- capacity,
- pricing and investment strategies, and finally
- capital circuits.

The third chapter describes a number of differences between the operation of large and small firms in construction. To account for these economic differences, it is necessary to consider the economic, social, political and technical characteristics of the business environment in which they operate. This is the subject matter of Chapters 1 and 2. These features together form the institutions of the social structures of accumulation, which specifically determine the way firms react in order to produce profits.

An important feature of the social structure of accumulation at any one time is the market in which firms conduct their business. The rules and regulations of every market are distinct, and in sum these rules and regulations comprise the economic institution of that market. Part II is concerned with construction markets and how they work in practice. Chapter 5 deals with the transaction cost approach applied specifically to construction, while Chapter 6 distinguishes between the different types of market to be found in construction. Each market is seen to have its own characteristics which impact on the methods of doing business. A universal supply and demand approach, with its emphasis only on price, is simply seen as inappropriate, when the response of firms to the nature of each market is described. Labour market processes are seen as operating completely differently from the processes of the market for property, which in turn is seen as quite distinct from the way transactions are carried out in construction contracting markets. The behaviour of firms in manufactured product markets, such as those supplying building components for construction projects, requires yet another description in order to be realistic.

In the first chapter in Part III, Chapter 7, we define *capacity* in terms applicable to the construction industry. In Chapter 8, the concept of capacity is seen in terms of individual firms. Firms are seen as having a basic drive to expand in order to remain competitive and survive. They are therefore continuously changing and developing, as they respond to new market and technological conditions.

These responses are considered in more detail in Part IV of the book, concerned with pricing and investment strategies. In Chapter 9, the pricing and investment decisions of firms are linked. Pricing decisions have implications for investment. The need to plan for growth requires investment funding, which needs to be consistent with the firms, ability to deliver profits of a sufficiently high level to allow a proportion of them to be put in reserve and accumulated for investment purposes. In the following chapter, the application of this model of strategic planning is applied to construction firms, who cannot make price and output decisions in the usual manner described in most economics texts.

In the final section of the book, we return to a broad view of the economic processes involved in the production of the built environment. In Part V we look at the overall nature of the system in which construction firms participate. Capital circuits are used as an approach which models both the use of resources and the distribution of the financial surpluses produced. In this way we return in the last part to the underlying causes of the unavoidable issue of inherent conflict to be found in any industrial setting. In Chapter 11 the industrial capital circuit is discussed, while the merchant, banking and property owning circuits are described in Chapter 12.

The aim of this book has been to give a clear understanding of some of the economic issues directly confronting construction firms in their day to day operations and to provide the economic basis for planning and decision making. We have also attempted to provide an explanation for the economic phenomena experienced every day by firms working in the construction sector.

STEPHEN L. GRUNEBERG
GRAHAM J. IVE

List of Acronyms

ACE	Advanced capitalist economy
APTC	Administrative, Professional, Technical, and Clerical
BS	Building Society
CAD/CAM	Computer aided design and computer aided management
CAP	Common Agricultural Policy
CFR	Construction Forecasting and Research
CIBSE	Chartered Institute of Building Services Engineers
CIC	Construction Industry Council
CIRIA	Construction Industry Research and Information Association
COP	Census of Production
DETR	Department of Environment, Transport and the Regions
DLO	Direct Labour Organisations
DOE	Department of the Environment
EMS	European Monetary System
ENR	Engineering News Record
EU	European Union
FAME	Financial Analysis Made Easy
GCS	Gross Capital Stock
GDFCF	Gross Domestic Fixed Capital Formation
GDP	Gross Domestic Product
GFCF	Gross Fixed Capital Formation
GOP	Gross Operating Profit
HA	Housing Association
HCS	Housing and Construction Statistics
ICE	Institution of Civil Engineers
ICOR	Incremental Capital Output Ratio
ISIC	International Standard Industrial Classification
IT	Information technology
JCT	Joint Contracts Tribunal
LA	Local Authority
LOSC	Labour only sub-contractor
MBO	Management-buy-out
MES	Minimum efficient scale

NACE (Rev1)	The General Industrial Classification of Economic Activities
NBER	National Bureau of Economic Research
NEDO	National Economic Development Office
NFCF	Net Fixed Capital Formation
NOP	Net operating profit
NPV	Net Present Value
NWRA	National Working Rules Agreement,
ONS	Office for National Statistics
O-O	Owner-occupiers
PBR	Payment by results
PFI	Private Finance Initiative
R&M	Repair and Maintenance
RIBA	Royal Institute of British Architects
ROCE	Return on Capital Employed
RPI	Retail Price Index
SET	Selective Employment Tax
SIC	The Standard Industrial Classification
SPC	Special Purpose Company
SSE	Social structure of the economy
SSHP	Social structure of housing provision
SSOCP	Social systems of organisation of the construction process
SSP	Social structure of production or provision
TGWU	Transport and General Workers Union
UCATT	Union of Construction Trades and Allied Technicians
VIBO	Vertically Integrated Building Owner
WES	Work to existing structures

Introduction

An explanation by way of introduction: the value of realism and the realism of value in economics.

The search for a theory of value in economics is for a 'unifying grand theory' capable, in principle, of explaining the totality of economic phenomena – a search for a common unit of measurement, to be sure, but beyond that, the underlying sufficient cause of values, in the plural (i.e. prices of this and that), and the force capable of yielding a determinate set of economic outcomes from certain non-economic givens – in short, the idea that both back economics, claims to be a science, by giving it a unifying object of study and a method, and at the same time lays it open to accusations of being no more nor less than metaphysics. For value is abstract, and not directly visible.

The great arguments in the history of economics have been arguments between theories of value (Cole *et al.*, 1991; Varoufakis, 1998). Following Cole *et al.* (and many others) we can call these contending theories of value the 'subjective preference', 'cost-of-production' and 'abstract labour' theories. Of these, the academically dominant school of thought, throughout the last century, has been based on the 'subjective preference' theory of value: rational, calculating, self-interested choice between given alternatives.

Economic debate has got lively and deep whenever this dominance has come under challenge – and 'economics' has stagnated into an orthodox body of doctrine whenever that challenge has faded. To be clear: we believe, along with an increasing number of other economists critical of the orthodoxy, that economics took a profound 'wrong turn' when it tied itself, as a 'reputable science', to the subjective preference theory of value in what is known as its 'neo-classical' form. We believe that this mistake has led both to bad theory and a wrong agenda for economics – and explains especially the difficulties which arise when it is attempted to apply economic theory to the interpretation (we will not say, explanation) of the 'real world' of modern industry. Nor, in our view, is it simply a matter of the 'wrong' theory of value having been chosen, whereas the 'right' choice would have solved all problems. Rather, one source of the difficulty for a practical economics lies in the drive, perhaps innate in all value-theories, towards an excess of abstraction, at too great a price in terms of realism and relevance.

However, statements such as those made in the previous paragraph open up such a range of arguments and issues between economists, and moreover arguments necessarily conducted in languages more or less impenetrable to non-economists, that it is certainly not our intention to develop systematically those statements or claims in the work that follows, in the sense of conducting an 'argument' or critique of orthodoxy – for to do so would exclude the possibility of this being a work of utility to students and practitioners in the production of the built environment. Instead, we hope the work will speak for itself, at least in respect of the range of issues our approach enables us to cover, the analytical methods we propose and the agenda of questions (for research and otherwise) which we raise. We do, however, think it both useful and necessary to give 'fair warning', especially to readers who have a knowledge of orthodox neo-classical economics – for such readers will not find in what follows much of what they will expect, and will find much that they will not expect.

We believe that our approach, as exemplified in the chapters that follow, is sufficiently consistent and straightforward as to be reasonably clear *in its application* to economists and non-economists alike. However, we also feel we owe a duty to both kinds of reader to explain in this introduction just what kind of economists we are – that is, what position we take in the fundamental debates that divide economists today.

The approach of many economists when they come to write 'practical' works on the economics of a particular sector or industry, is to start with received value theory, expound a version (to a greater or lesser degree simplified) of that theory, and then to give 'examples' of its application by nominally substituting apparently recognisable phenomena of that industry into the purely formal categories of that theory. An instance would be a discussion of market price for a commodity in terms of the thought-experiment that demonstrates the possibility of a set of pre-reconciled independent choices made by possessors of given productive endowments and given consumption tastes (each with perfect knowledge, and all acting in 'analytic' time to explore all hypothetical options fully before making actual choices) leading to an equilibrium of demand and supply; moves on to describe this imagined market as one of 'perfect competition'; and then discusses how the idea of perfect competition can be used, to some extent or other, to describe and illuminate actual markets for grain, or fish, or stocks and shares.

Our approach, by contrast, is to start by making a model of the actual *processes* by which economic actors arrive at their decisions – in terms which we hope will be recognised by practitioners as capturing certain (inevitably, not all) interesting or significant aspects of that process. We are in a position, since we are concerned here only with the construction industry, to develop and to prefer 'local' or special to general theory – to adapt our models to capture local circumstances, even at the cost of loss of ability to generalise about the economy at large. This we are happy to do.

The substantive content of our theory is to an extent eclectic, formed by taking and melding together, magpie-like, whatever catches our interest from a diverse range of sources. However, just as magpies prefer that which glitters from the array open to them, so we prefer that which looks to us more 'realistic'. Our approach often involves simply relaxing the stays of a pre-existing theoretical corset, by introducing some added degrees of realism.

All of this courts the danger of over-compensating, to the extent that any and all classificatory or descriptive coherence is lost – the point where each instance becomes unique. Naturally, we hope that our readers will agree with us that we have not gone so far as that.

One powerful inspiration for us, in our quest for a sufficient, minimal consistency and theoretical coherence, has been the approach to the treatment of time, uncertainty, surprise, the past and the future. Time, throughout what follows, is we hope almost always *perspective* not *analytic* time. In perspective time, decisions are made sequentially, not simultaneously, and are made using the exercise of economic imagination about possible futures – that is, under real uncertainty, where what is envisioned *ex ante* often differs from what is realised *ex post*. In this, like other economists of construction (Hillebrandt, 1985; Bon, 1989) we have been inspired by the writings of G.L.S. Shackle and other 'Austrian' economists – even where we disagree sharply with the political economy of much of Austrian economics. It is to Shackle, however, that we owe a particular debt – and Shackle was both an 'Austrian' and a post-Keynesian.

The post-Keynesian economists are, of course, best known for their macroeconomics – and for insisting upon the profound theoretical implications of Keynes' critique of macroeconomic orthodoxy, and thus resisting the absorption of the Keynesian legacy into a slightly modified orthodoxy – the so-called Keynesian neo-classical synthesis. However, there is also a very distinctive and coherent post-Keynesian microeconomics, and this has often provided us with at least a starting point or point of reference. We are also indebted to those economists who have sought to fuse elements of Keynesian and Marxian economics – the tradition beginning with Kalecki and Robinson, and carried on today by, *inter alia*, such inspirations for parts of our work as Bowles, Weisskopf and Marglin.

Nearly twenty years ago, a book appeared whose publication in our view (and that of many) deserves to be seen as seminal to the development of a truly modern, non-orthodox economics – Nelson and Winter's *An Evolutionary Theory of Economic Change* (1982). The 'Introduction' to that work contains an argument about the nature of the malaise afflicting orthodox microeconomic theory (what is known to economists as general equilibrium theory), and directions for its remedy, with which we wholly agree. Indeed we could wish, if academic convention and copyright law did not forbid us, to reproduce it more or less entire as our own introduction to the present work. Among the very many indications therein for positive directions for a new microeconomics let us cite at least the following:

(1) theory must seek to comprehend, in stylised settings, the unfolding of economic events over (perspective) time; we must escape from the grasp of a purely 'analytic' time, that is really no time at all

(2) firms are motivated by profit, seek it and search for ways to increase it, but 'firms' actions will not be assumed to be profit-maximising over well-defined and exogenously given choice sets' (p. 4)

(3) analysis should not focus on hypothetical states of industry equilibrium, in which all unprofitable firms have left the industry and the profitable ones are at their desired size

(4) firms learn; at any given time a firm has limited capabilities, and habitually uses certain decision-rules: 'Over time these ... are modified as a result of both deliberate problem solving efforts and random events'

Meanwhile, orthodoxy itself has been invaded by the practitioners of the so-called 'new institutional economics', an economics of incomplete information, bounded rationality, complex organisations and transaction and organisational costs. This we welcome, and use happily where the only developed alternative is the old orthodoxy – while recognising that it too is suffering the fate of Keynesian economics – to be re-absorbed, in somewhat travestied form, into a revised orthodoxy.

This, then, is the kind of economics, and the approach to the use of economics to study industry and business, that a reader will find exemplified in what follows. We hope you, the reader, will find it yields both light and fruit.

Part I
Social Structures of Accumulation

1
Social Structures of Organisation

Introduction

This chapter introduces the concept of social structures applied to various aspects of the organisation of the construction process. These social structures relate to the *economic institutions* which determine the working arrangements for undertaking construction.

These arrangements are made between people working in organisations within *rules* and *routines* which create their immediate economic environment. Firms are at the active centre of the analysis, as the actors taking decisions – but are seen as responding to their economic environment, the conditions in which they must carry out their transactions. Moreover, these conditions not only determine how firms relate to other firms, but also how firms organise themselves internally.

As capitalist firms their purpose is ultimately to generate sufficient profits to accumulate capital. This chapter seeks to describe the overarching social structures which form the context for economic decision making by firms. The next chapter develops these ideas of social structures by using speculative housebuilding and the housing market as an example.

Social economics: an introduction

The primary institutions of the economy are markets and firms (for, as we shall show, markets too are institutions), with public institutions pervasively in the background, holding the ring and, at least, regulating these other institutions. The production of commodities by firms, and their sale in markets, requires and draws upon a set of technical

3

and social arrangements of production itself (the labour process and technology at the heart of the economy). There are also a number of institutions (which we call systems of provision) specific to particular sectors of production. These systems of provision guide and direct the organisation of production in terms of flows of money and commodities. Finally, production also draws upon four overarching sets of economic institutions (see below) as well as institutions belonging to the social and political domains.

These four sets of overarching economic institutions are those of the systems of provision of labour, of land and infrastructure, of money and of information or knowledge. These are the ubiquitous prerequisites of firms' production of commodities not themselves producible by firms as commodities. They are distinguished from other non-economic (legal, political) preconditions for capitalist production by being directly embedded, integral parts of the process of production and circulation of commodities, as providing what economists call *inputs*. They have a quantitative as well as qualitative dimension.

To these we must add the social institutions that embody, in 'externalised' form, the social norms of constraints on the behaviour of each individual in that particular society; and the political institutions that embody the extent of and access to the power of the state. This obviously includes, but is not confined to, the legal system, and the particular forms of private property rights.

All these, together with the institutionalised markets in produced commodities, constitute the *environment* within which individual capitalists (the controllers of firms) form their expectations and make their decisions about production and investment.

> [T]he accumulation of capital through capitalist production cannot take place either in a vacuum or in chaos. Capitalists cannot and will not invest in production unless they are able to make reasonably determinate calculations about their expected rates of return ... Unfortunately, ... although many economists may recognise the importance of external factors, most have nonetheless left the investigation of those factors to sociologists and political scientists.
>
> We argue, in sharp contrast, that macrodynamic analyses should begin with the political-economic environment affecting individual capitalists' possibilities for capital accumulation ...We refer to this external environment as the social structure of accumulation.
>
> (Gordon *et al.*, 1994, p. 13)

We concur but, since our purpose is not macrodynamics but a purely 'local' analysis of one sector of the economy, we divide our account of the relevant environment into the general social structure of the economy (SSE) and the local social structure of the organisation of the construction process.

General and sectoral structures: social structures of the economy and social structures of a branch of production

It is one thing to speak of a social economics, and to say that economic activity is socially conditioned. It is something rather stronger and more specific to speak of a social *structure* of the economy or of a branch of production.

The most important micro-elements giving stability to the social structure of the economy, and thus enabling us to speak of a *structure*, are enduring institutions. We must therefore consider where such institutions come from and how they persist and change. The *routines* of individuals through repetition become externalised and confront the individuals as the norms and customs of institutions (such as firms) formed by this self-same stability of individual behaviour and of interpersonal relations. Stability of institutions therefore permits individuals to form both stable habits and stable expectations of the future.

These stable institutions are to be found at both the sectoral or immediate level and at the national or aggregate level. We shall call a set of stable aggregate or overarching institutions, pervasive across a whole national economy, a *social structure of the economy* or SSE (see papers by Gordon *et al.*, Reich, and Weisskopf, all in Kotz *et al.* (eds), 1994). Whereas, a particular sub-set of stable markets, firms and individuals concerned with producing and transacting one product, such as new housing, we will call a *social structure of provision*. For example, in this and the next chapter we discuss the social structure of housing provision, or SSHP for short. To complete the picture, we shall need the term *social structure of production* to refer to the labour process – the social organisation of the productive work – which produces the physical goods and services that become the content of the flows of commodities, measured by their money values, between transactors (developers, builders, building owners, building users) in what we have called the process of provision.

The focus in Part I of this book is mainly on the social structures of provision. For equally full discussions of the social structure of the economy and of the social structure of construction production we must refer the reader elsewhere. However, some brief remarks on these are indispensable.

The social structure of the economy and its overarching institutions

The seven key social relationships within the social structure of the economy are:

1. between economic classes: i.e. between capital and labour (labour market, real wages)
2. between firms in same industry (national competition; forms of competition; regulation of competition)
3. between firms in different branches of the economy (terms of trade, i.e. ratios of prices of produced inputs and outputs; subsidies and indirect taxes; the capture of economic rents)
4. between national and foreign firms (international competition and international terms of trade; international value of national money, exchange rates)
5. between debtors and creditors, industrial and financial firms (purchasing power value of money, i.e. inflation; liquidity of other financial assets)
6. between industrial firms and landowners or property capital (prices of land and natural resources)
7. between producers, users and holders of economically useful knowledge

These relationships all potentially involve conflicts of economic interests.

Of course conflict is not all of social reality. Clearly all members of a society do have common interests. In particular, all can potentially gain (which may even translate into: 'are more likely actually to gain rather than lose') from anything which *costlessly* facilitates more efficient production, or prevents destruction of the social infrastructure. Yet these shared interests are hardly sufficient to shape action and policy. Pure win–win solutions to problems do exist (though even then potential exists for dispute over entitlements to larger shares in the division of gains). Much more common of course is the change which will not satisfy Pareto's criterion, of leaving some better off and none worse off.

The key institutions are therefore those which have emerged, in a particular society at a particular period, to attempt to regulate or settle the outcomes of actual and potential conflicts in each of the above set of social relationships. These conflicts arise out of conflicts of interest and the existence of other institutions formed to express and promote those interests. The outcomes may represent the consolidation of the victory of a particular interest, or coalition of interests, or may represent a compromise between undefeated antagonists.

Those key institutions that relate to intra-national conflicts of interest normally, but not always, are aspects of the role of the state. In turn, that role of the state may be either to mediate compromises or to aid and represent particular interests at the expense of others, or it may contain elements of both roles. The role of the state is in fact equally decisive regardless of whether it takes the form of obvious state intervention or a less obvious resolve not to intervene. For example, relations between landowning and other interests are always a matter resolved by the state – whether that resolution takes the form of policies of land value taxation, land nationalisation, a panoply of land-use planning controls, or their opposite.

Inter-national settlements of conflicts of interest, in contrast, are only by exception within the role of the nation state. Some national states at some time have had the power to become internationally strong states, capable of projecting their power abroad, diplomatically and, ultimately, militarily, without meeting effective resistance. Examples are Britain for part of the nineteenth century and the USA more recently.

International economic institutions and the UK economy

The discussion in this section relates to item 4 in the list of key social relationships (see p. 6).

One dimension of any national SSE, therefore, is this: is there a single internationally dominant (hegemonic) state dominating the international economic relations of the economy in question; and if so, which state is it, and how does it express its interests? Such dominance is expressed, above all, through the international monetary system, and also through the international trade system. Up until the 1970s, the USA was fairly clearly hegemonic with respect to Western Europe (and particularly, Britain) in matters of inter-national economic arrangements. The relevant system comprised the institutions of the post-war reconstruction period (the US-dominated IMF, the dollar as international currency of international financial settlements, fixed dollar-gold exchange rate and dollar-gold convertibility, etc.) (Maddison, 1991; Armstrong *et al.*, 1991).

After the dollar crisis of the early 1970s, and more especially after the rise in the economic power of the EU and Japan relative to that of the US, it has been argued, the US ceased to be economically hegemonic, and new, more multi-lateral, but also less settled and less stable arrangements of international economic conflicts within the advanced capitalist group of countries took its place (Brenner, 1998). At the same time, emerging pressure from international banks (especially but not only US-based ones) for unrestricted mobility of financial capital resulted in most countries eliminating the controls on international capital movements that almost all had maintained up until 1980. Among other things, this it is argued had the effect of greatly increasing the instability of exchange rates. For, post-1970, each currency had a multiplicity of meaningful exchange rates (not only that with the US dollar); and when currencies began to float against each other this greatly increased the uncertainty of return on any investment with an international dimension to its financing, costs or revenues.

International trade was from the post-war years controlled by GATT (General Agreement on Tariffs and Trade) in a regime aimed (successfully) at achieving tariff reduction in trade between all the advanced capitalist economies. Bilateral US negotiation of most-favoured nation terms and of openings for the turn by US industrial corporations towards becoming multi-national producers, were equally important in the stable post-war period that ran until the 1970s. For the UK, this was however a stability of relative decline, as the UK's share of rapidly-growing world trade steadily fell, and deficits in the balance of trade in goods became chronic.

In the 1980s and 1990s, the decisive changes to the international trade regime (for the UK) were: the formation of the Single European Market, and the relative decline of UK trade with the rest of the world; a growing dependence on UK exports of services; and transmutation of former chronic balance-of-trade crises into exchange rate/interest rate crises.

National economic institutions of the UK economy: the post-war period and the current period

The discussion that follows relates to items 1, 2, 3, 5, 6 and 7 of the key social relationships within the social structure of the economy (see p. 6).

In the period from 1945 until the 1970s (postwar period) and then in the period from the 1980s to the present (current period) we find, in very summary outline, the following:

(1) Labour and the distribution of income between labour and capital

Postwar period

Systematic attempts at state 'mediation' of capital/labour conflicts, both over real wages (and thus wages as a share in output relative to profits) and over changes in the labour process (and thus productivity). These by the 1960s had fused into state policies which attempted to link wages with both prices and productivity (Prices and Incomes Policies). The aim of the state was to support faster growth of productivity in industry by linking both real profits and real wages to the achieved rate of productivity growth; whilst reducing open outbreaks of industrial conflict (strikes) by converting bilateral (employers' associations, unions) industrial collective bargaining into tri-lateral (including government) 'social contracts', by including in the scope of bargaining levels of taxation of incomes, levels of state expenditure and state price-controls. The state's 'ideal' was an orderly and agreed division of the benefits from increased productivity within the society as a whole. This broke down under pressures from slowdown in the actually-achieved rate of productivity growth, increase in the share of output claimed by the state for redistribution and for public services, rigidities in socially acceptable income differentials, and the absence of consensus over any 'agreed' division of incomes, all in a context of accelerating inflation and thus unforeseen changes in the 'real' consequences of agreed 'nominal' divisions of income.

Current period

The neo-conservative government of the period beginning in 1979 consciously exerted state power to break the pre-existing stand-off between organised labour and capital in the interests of capital, and both by legislation and by its role in a series of political-industrial conflicts overwhelmingly succeeded.

The position of the state became that income distribution should be the result of individual (not collective) bargaining over incomes, and that the results of this bargaining would reflect balances of market supply and demand for the 'factor' in question – each should and could 'charge what the market would bear' for whatever particular skill, etc. they were providing. The institutions of national collective bargaining were largely dismantled, and those of the 'social contract' completely so.

The existence of high unemployment was taken to indicate that labour markets were not clearing because wages had not yet fallen far enough, for particular categories of labour. 'Rigidities' in labour markets (i.e. anything which stopped wages falling to find the 'market clearing' level) were the target of action by the state. Labour market policy came to mean a series of measures to reduce such rigidities. New labour market institutions were created to reduce rigidities and increase flexibility – including substantial reductions in the system of state benefits for the unemployed, and new arrangements for wage setting in the public sector. However, labour market policy remained linked to policy on the price level (inflation) in that a return to pre-war 'Treasury orthodoxy' indicated that the way to reduce inflationary pressure originating in the labour market was (until such time as such pressures could be eliminated by attacks on 'rigidities') to increase unemployment, and thus reduce the bargaining power of workers, and force nominal wage restraint.

The goal of policy became to run the economy with demand at a level which ensured there was sufficient unemployment to control wage pressure on inflation, and at the same time to raise the share of profits in national income – goals broadly achieved (with the serious exception of the end-years of the late 1980s economic boom in the UK).

The period also saw fundamental changes in the system of training of new workers, including the collapse of almost all bilaterally-run (i.e. jointly by employers and unions) training and apprenticeship schemes, and a continuing series of experiments with new training institutions.

Transition to a New Labour government in the late 1990s has not significantly altered this picture (except that macro-demand management, in the form of control of interest rates, has been passed from the Treasury to the 'independent' Bank of England – a key 'institutional' change).

(2) Competition and market structure

Post-war period

- Competition in UK markets for manufactures was still mostly national, but steadily becoming more international, while an increasing share of UK manufacturing became taken by local branches of US multinationals.

- Concentration ratios in UK industry increased. Government policy increasingly favoured concentration, as leading to efficiency gains and formation of 'national champions' able to compete internationally.

Current period

- Competition in UK markets for manufactures is now overwhelmingly international. The increase in international competition has spread to some service industries.
- Multinational penetration of UK production spread to include Japanese and European multinationals, and spread from manufacturing into services.
- Specialist financial institutions developed strongly whose role is to facilitate mergers and acquisitions, but also demergers (especially 'management-buy-outs' or MBOs).
- Concentration ratios in terms of UK production continue to rise, but more slowly.
- Concentration ratios in terms of UK market shares fall, as a result of increased international competition.
- Cross-boundary mergers and joint ventures become important.

(3) Terms of inter-sectoral trade

Post-war period

- Subsidies to agricultural producers and to some nationalised industries permit lower prices for their products.
- There is an uneven structure of indirect taxation (high rates on some goods, zero rates on others).
- Some attempts are made to make the tax system favour manufacturing industries (e.g. Selective Employment Tax; investment tax reliefs and investment grants).
- Producer power enables seller-industries to capture economic rents as 'producers' surplus'.

Current period

- The Common Agricultural Policy (CAP) replaces subsidies by guaranteed prices. Nationalised industries are released from price restraint and privatised. There is a change to more uniform structure of indirect taxation (VAT).

- The purchasing power of large buyers increasingly enables customer-industries to capture economic rents as 'consumers' surplus'.

(4) International trade and financial system (see p. 7)

(5) The national financial system and the value of money

Post-war period

Monetary policy in this period meant control (by Treasury and Bank of England) of the money supply, the supply of credit and interest rates. By the end of the post-war period, the domestic purchasing-power value of sterling had become subject to continuous erosion, at rates that both oscillated through each economic cycle but also 'ratcheted' upwards from one cycle to the next, and ended in both stagflation (inflation failing to fall even when the 'real' economy was in recession) and fears amongst some that an 'explosive' inflation had become a possibility.

Current period

The start of the current period saw a brief and unsuccessful experiment with institutions of a 'monetarist' regime – i.e. explicit state monitoring and control of the money supply at a rate intended to control inflation. Since the end of the late 1980s boom, inflation has been reduced to low and stable rates. The long term expectations held regarding future inflation can be deduced from the historically low present (1999) level of 'real' long-term interest rates in the UK.

Some have begun to worry, indeed, that there is now a real possibility of a major deflation (fall in general price level), especially, an asset-price deflation (Bootle, 1996).

(6) System of provision of 'development land' and of natural resources

The system of provision of development land is discussed below, and at length, as part of 'Social structures of housing provision', and is therefore not discussed here.

Post-war period

The period began with the nationalisation of coal mineral rights. Government adopted a policy to encourage national self-sufficiency in agricultural products as far as possible (for military/strategic reasons). The main fuel source of the economy shifted in this period from coal to petroleum oil, led by a regime of 'cheap oil' (imported). Ownership of overseas oil mineral rights lay initially in the hands of US and UK oil multinationals. Over the period, these mineral rights were often nationalised (e.g. in the Middle East) and governments of oil-producer economies came to regulate extraction rates. This led eventually to the explosive increase in crude oil prices in 1973 and subsequently, and to the 'oil crisis'.

Current period

Fears of growing shortages of key natural resources, especially oil, led to systemic 'counter-measures': policies of energy-efficiency; development of alternative technologies and of substitutes; opening of new sources of supply previously regarded as 'marginal' (too expensive to develop). The UK became a major oil producer. Terms-of-trade prices of most natural resources fell by 1990s to historically low levels.

The CAP regime continued, but with reductions in guaranteed price levels and with some 'volume' restrictions on production introduced. Prices of most agricultural products fell to low levels by the 1990s.

(7) System of provision of existing and new knowledge

The first requirement of a system of knowledge provision is the reproduction, or transmission between generations of people, of the stock of existing technical knowledge. This is done partly inside firms, and partly in specialist institutions of education and training. Dynamic accounts of systems of knowledge provision must deal also with technological progress, the addition of new possibilities to the social (total) production set – i.e. to the set of production possibilities the elements of which are held by at least one firm. Either this new knowledge is produced exogenously (outside of the world of firms – for example, developed in universities and government research institutes and published in scientific and technical publications), and simply added to the stock of public information; or, firms engage in private research and development activity (R&D) as a kind of investment activity, whose value or payoff lies in its potential to expand the production set of the firm undertaking it (the knowledge thus produced either remaining private or, if published, protected by patents).

The whole complex sphere of the social structure of technical knowledge has here had to be simplified and reduced to one or two 'stylised facts'.

Post-war period

Technical knowledge is in part public and shared by all firms, and in part private and untransferable; new technical knowledge is both exogenously generated (outside of the world of firms) and endogenously produced as the output of a distinct activity called 'research and development'. This is the age of co-existence of major government industrial research laboratories and (growing) in-house corporate R&D departments.

Technical knowledge in some cases exists in the form of 'units', each sharply bounded, articulated and complete – i.e. all that is necessary in order to use a certain technique. Each of these 'units' is the in-house technology of highly integrated firms. Technical knowledge and capability is still relatively cheap for a firm to acquire. Technical knowledge is increasingly separable from the other inputs used by a firm – in particular, separable from 'ordinary' labour, which is therefore treated as basically homogeneous, so that most workers are treated as unskilled ('semi-skilled'), and highly transferable between industries or techniques. The occupational categories of *technologist* (or applied scientist) and *technician* become important, and with them, technological higher education institutions, for the education and training of such staff. Craft-based knowledge, transmitted through craft apprenticeships and, less formally, between experienced and inexperienced members of crafts, declines in relative importance, but remains significant.

Much technical knowledge remains a matter of 'know-how', which in turn is largely a matter of practical experience. Most know-how remains implicit and not articulated, still less written in manuals. Reproduction of know-how continues to be most successfully organised through the craft system.

Much *new* technology, in UK, consists in this period of lagged adoption of techniques first developed in the USA, and already in industrial use there by the 1940s.

Current period

Firms that have acquired their own strong institutionalised systems of R&D and advanced training of technologists are among those that best survive the general 'British deindustrialisation' of the 1980s and 1990s (e.g. chemical and pharmaceutical industry). Other survivors include firms who form *technology partnerships* with government and research institutes, enabling them to utililise the latter's work on initial development of new technologies.

The craft system of reproduction of know-how largely collapses, severely weakening many firms dependent upon it for their know-how.

Successful knowledge-based firms, employing mainly highly qualified professional and technical staff, develop methods of acquiring and retaining significant levels of 'private' technical knowledge in the process of carrying out their 'ordinary' activities, i.e. not as 'output' of separate R&D departments.

Technological higher education expands and these institutions move into mass education of technicians.

So-called 'knowledge-products' acquire large markets (e.g. computer software).

Technological networks between sets of firms become important, in which no one firm possesses all the technology needed for a 'complex system' product.

Information technology is the single most important new technology, more pervasive across the economy than any single new technology of the post-war period.

The cost of acquisition of a new technology for a firm becomes very high. The accelerated rate of technological change tends to make 'imitation', using 'public' knowledge, less feasible as a competitive strategy.

(8) Overall structure of price expectations

Firms in UK, therefore, including both construction firms and their corporate customers, made their output and investment decisions in the context of the following set of expectations.

Post-war period

The general price level was expected to rise slowly and steadily.

Prices of key natural resources were expected to be stable in 'real' terms, at historically low levels, i.e. to stay within a narrow range relative to existing 'real' levels.

Real wages were expected to rise steadily for all types of labour, and faster for workers in the 'primary' labour market (Ive and Gruneberg, 2000, ch. 2).

Real interest rates were expected to stay low, but remain positive.

Property prices (offices, housing) were expected to rise slowly in real terms.

Land prices for development land expected to rise at similar rates to property prices.

In this period large firms, who came to control a rapidly increasing share in UK output, increasingly engaged in long term planning and forecasting; and did so with confidence that the 'single value' forecasts they used for key variables would not prove too far from the truth (though UK firms' use of long-term forecasts was vitiated in some sectors by uncertainty over the short-run business cycle).

Many of these confident expectations were 'falsified' by events as they unfolded in the 1970s.

Current period

Monetary variables (exchange rates; inflation rate; nominal and real interest rates), in the aftermath of the 'shocks to expectations' of the 1970s, have been expected to possess a wider range of possible future values. In the late 1990s eventually firms became prepared to focus on a narrower range of possible values for these variables.

Real wages came to be expected to depend upon 'local' (i.e. specific) supply/demand conditions, and thus to show different rates of change for different types of labour (expectation of a 'mix' of labour surpluses and shortages).

After the property price boom of the 1980s, expectations in 1990s have been for property prices not to increase in real terms.

In this period large firms have lost much of the confidence they used to have in 'single value' forecasting. Most have switched to some form of 'scenario planning', in which focus is on a number (usually two) of widely variant forecasts about the context in which the firm expects that it is possible that it will find itself (Loasby, 1976; Drucker, 1994 and 1995).

Key elements in the social structures of the economy: structure of this book

The implications of (1) labour and capital relations for the construction industry (its firms) are dealt with briefly in Chapter 3 of this book, and in parts of the companion work (Ive and Gruneberg, 2000) – a partial omission from this volume forced by lack of space. The implications of (2) and (3) (national competition and terms of trade) for the construction industry (its firms) are part of the subject matter of Part II of this book.

The implications for construction firms of (4) (international competition and terms of trade) are only dealt with briefly, in Chapter 7 of this book. This relative brevity is because production of the built environment is still only very partially internationalised. We recognise that it would require a larger treatment in a truly full account.

The implications of (5) (the financial system) and (6) (the provision of land and natural resources) are part of the subject matter of Part V of this book. The implications of (7) (the knowledge system) are not covered in later chapters – an omission forced by both our lack of expertise and lack of space. The production, pricing and investment decisions of the firm itself, to which all this is the context or environment, are the subject matter of Part IV of this book.

Social structure of construction production: technology and labour process

We will consider a social structure of construction production under three headings – the social structure of technical knowledge, the social structure of control of the labour process, and the social division of labour. Social structures of production are made complex by the differential application of technology and forms of control as between different sectors of the economy. Thus for example, even at the height of the Fordist SSP (social structure of production), not all sectors of production were dominated by 'assembly-line' technologies or forms of control. Thus we will make only some brief introductory remarks about general, economy-wide SSPs, and concentrate instead upon 'local' (i.e. construction-sector-specific) forms of those structures.

The standard schema put forward by proponents of general social structures of production (Lipietz, 1987) identifies three such structures in the twentieth century: first, the Taylorist SSP, based upon 'scientific management' and (managerial) methods of task analysis and simplification (such as 'work study'), and recomposition of tasks by managerial planning and pre-calculation; next, the 'Fordist' SSP, based upon mass production technologies, sequences of tasks linked by conveyor belt or assembly line, and 'machine-pacing' of human work; and last the 'post-Fordist' or 'flexible' SSP, based upon application of information technology to production, CAD/CAM, numerically-controlled programmable machines, automated plant, sequences of machinery, with human work reduced to instructing, programming, monitoring and servicing the machinery (see Ive and Gruneberg, 2000, ch. 2, for references and a somewhat fuller account).

The construction sector, however, has throughout possessed its own, somewhat 'peculiar', institutions of technical knowledge, labour-process control and occupational structure.

Technical knowledge in construction

The technical knowledge base of firms in the construction sector is not for the most part structured around in-house R&D departments, or the patented 'outputs' thereof. Instead, more than in most industries, it is a matter of inarticulable and implicit knowledge, dispersed amongst many employees of the firm, but nevertheless knowledge 'of the firm', in that it consists of knowledge of the firm's repertoires and routines for handling a diverse range of operations, and for responding to an equally diverse range of technical challenges or problems (Groak, 1992).

The engineers or technologists employed by the firms which have the role of co-ordinating the production process (main or management contractors) hold only part of the knowledge required for a complete production process. Other parts are held by designers employed by separate professional firms, by the engineering staff (and indeed site workers) of specialist subcontractors, and by the staff of component manufacturers. The production process operates from day to day not only through comprehensive and sufficient sets of written instructions to be followed, but also by relying upon the experienced judgement of site workers.

Although still organised in large part through professional institutions originally committed to the idea of autonomous competence of an independent individual member of their profession, in practice it is recognised that it is increasingly unlikely that one individual can possess all the technical knowledge required for 'competence' in this sense – which therefore becomes a matter of the competence of firms or of networks of individual professional specialists.

Whether because of high levels of inter-firm labour mobility, or because of the effects of the system of labour only subcontracting (see below), or for other reasons, technical training of the operative work force has for a long time been organised at the level of the industry (Construction Industry Training Board), and not the individual firm.

Control in the labour process

A labour process is the carrying out of a particular kind of work, a transformation, by the application of concrete human skills and work capacities, of specific natural resources into equally specific finished products, ready for human use. In industrial capitalism generally, the production of such a final product will typically be highly 'indirect', involving the production of many 'lower order' goods each of which enters (as a purchased commodity) as an input into other stages of the process. These 'lower order' goods can be divided usefully into the *instruments* (tools, machines, etc.) and *objects* (materials) of subsequent parts of the labour process.

A labour process involves applying human effort to transform objects, with the aid of instruments, into useful products. It includes conception (what exactly is to be produced, how, by whom) as well as execution.

Control issues arise once the labour process as a whole is not the direct domain of the producers themselves, and alone – either as individuals or as simple co-operators. Formal control passed long ago to employers. This control was deepened when managers acquired the technical capability to organise the conception of the work to be done – to design tasks and issue technical instructions, and to devise large-scale processes dependent on elaborate and expensive machinery (see Ive and Gruneberg, 2000, ch. 2).

Key construction sector institutions of labour process control include: a high level of development of systems of communicating complex instructions as drawings; and use of relatively large numbers of site supervisors. On the other hand, payment-by-output or PBR (payment by results) wage systems have developed strongly and persisted in construction precisely, in part, to avoid and circumvent need for direct and continual supervision – and sometimes to make one of the production workers themselves perform the role of supervisor (as with gang-based PBR).

Labour only subcontracting (LOSC) has become the dominant form of employment in the UK construction industry. This accommodates various 'indirect' forms of work control, and places reliance on immediate incentive payment structures to control productivity.

Project management techniques developed in recent decades assist managers in monitoring the performance of sets of workers against project-targets, for time as well as cost.

Uncertainty for firms is endemic to any labour process. However, in construction labour processes, as currently organised, the uncertainty is mitigated or transferred to others in particular ways.

Actual production (as opposed to the potential production set that is the result of technical knowledge) is a process in which a set of physical potentialities for transforming inputs into outputs, a set given by the state of technical knowledge and limits on supplies of specific kinds of inputs, is both *narrowed* into a specific technique and made *actual* – that is, *ex ante* imaginings of possible outcomes (which stimulate firms to attempt production) are succeeded *ex post* by results, in terms of accountable quantities of inputs used and outputs actually produced. The productivity of a technique is not (cannot be) guaranteed in advance. The imagined productivity that stimulates the capitalist to risk their capital in a particular production activity may either have been conceived as a 'best case' possibility (as in a technical manual for a machine, where 'rated' output per period of time represents the best output feasible if there are no interruptions to or slowdowns in its operation) or as a 'most likely' ('mode') or 'mean' outcome (based on observation of a large number of similar past operations carried out in assumedly similar conditions).

Most especially what is intrinsically of uncertain productivity is the human labour (work, effort) hired for the production activity. A firm purchases labour power, the potential to work for a certain period of time. How much work is actually done in that time, and with what effectiveness, remains to be seen at the time the labour is hired. It may be 'possible' for a gang of workers, using a specified set of machinery, to pour x amount of concrete in an hour. But it is also possible that it will transpire that, in any given hour, no concrete was poured (say, because the flow of supply of materials had broken down; or, because of weather conditions; or a strike; or because of rejection of test samples of output from previous work); or that some amount between zero and x was produced.

Some of these possibilities may be regarded by the firm as less 'likely' than others (meaning, the manager in question would experience either more or less 'surprise' should they actually transpire) (Shackle, 1968, 1970). The firm may even believe that, if it analyses production in terms of small enough units of activity, each 'repeated' many times within the whole, it can apply probability distributions to outcomes, based on data from past experience.

However, the institutions of construction subcontracting (including a certain context of inequality of bargaining power between contractor and subcontractor, whether the subcontractor be a small firm or a LOSC gang) have enabled main contractors to transfer much of this uncertainty onto others. With subcontracting, all that is needed for main contractor certainty is the ability to form and enforce (implicit or explicit) 'lump sum, all risk' subcontracts. Once such contracts are placed, production costs then become more or less predictable for the main contractor.

The social division of labour

Specific or concrete labour processes, as discussed above, are put together and (to a degree) controlled by managers of a firm. But they are put together by combining and using socially-generated elements and relationships. For example, 'labour' is differentiated for practical purposes into different 'types', and requisite amounts of each are hired and engaged in the specific labour process. But these labour-type categories are social constructs, generated and sustained outside of the specific workplace.

The number of 'types' and the definition of each type ('crafts', 'trades', 'skills', 'occupations') are a crucial dimension of a social structure of production. The social system of recognition of occupations may be very strong – as with craft and professional definitions of occupations as essentially readily 'portable' by individuals between firms – or less so, as with occupations defined within internal labour markets. In construction most firms still take the occupational categories of the employees they hire as socially given, and thus reduce their employment requirements to quantitative ones (i.e. how many workers, of each given type, do they need).

Innovation in the labour process

Most firms, most of the time, simply use the socially prevailing forms of labour process control. However, from time to time firms innovate in this area, by introducing new definitions of jobs, new process technologies, and new forms of control. Insofar as these innovations succeed then, just as with any other innovation, the forces of competition between firms are likely to ensure their spread/diffusion.

However, the first key for firms in terms of their planning of production and investment is the extent to which the 'given' SSP gives them stable, predictable production outcomes, in terms of time, quality and cost. Whereas, under the post-war SSP in construction, production costs became liable to take any of a range of possible outcomes, we argue that in the current SSP this uncertainty of possible outcome has been significantly reduced.

Introduction to concept of Social Structures of Provision: routines, conscious decision-making and expectations

The primary institutions of the production of commodities are firms. Only individuals, in the strict sense, can be said to act purposively, insofar as they may have an aim in mind. Nevertheless boards or committees, or assemblies of voters, as well as individuals, can take decisions.

According to Williamson (1975, 1985) and the school of transaction cost economics, economic institutions arise as solutions to problems of information. We do not disagree, but rather wish to add the recognition of the importance of habit in individual, and routine in social, life, and of social norms. In brief, we consider individuals as social beings, as formed in society and *socialised*, rather than as pre-given bundles of tastes, talents and endowments who find it advantageous, by criteria of strict calculation of rational self-interest, to invent society (social institutions).

Firms attempt to make rational, planned decisions. But the domain of rational, maximising decision making is, even within firms, a limited one. The social life of individuals, including their life as members of firms, is only possible because of the existence of routines which avoid what would otherwise be constant information and stress overload.

Habitual behaviour in part represents the continuation of, or consequence of, actions once consciously considered and chosen. But only in part, for many such behaviours are acquired by imitation, and by adjusting to pre-existing norms and expectations. Routine or habitual behaviour by individuals interacting with one another may be said to give rise to social institutions which embody these routines as norms and rules. But equally, the institutions an individual experiences may be said to shape the purposes, aims, values and preferences of the individual. Thus the individual and the institution are chicken and egg. This is where we differ from transaction cost economists such as Williamson for whom, 'in the beginning is the individual', from whom first the evolution of the market and then, later, the firm can be explained (using the approach known as methodological individualism).

Institutions therefore involve routinisation of the behaviour of the individuals affected by them. Thus, to illustrate with reference to the firm:

- the behaviour of a firm becomes largely predictable regardless of the particular individual acting in its behalf (Weber's bureaucratisation);
- individuals working in, or in transaction with, a firm pick up clues and models from these institutions to guide their behaviour, in the sense of learning behaviour and skills.

Nevertheless, from time to time economic individuals do make consciously considered decisions, or formulate new plans. The content of the decisions thus taken is not 'determined' by the context in which the decision-makers find themselves. Nelson and Winter (1982) contains a brilliant discussion of the interrelation of routine and conscious decision-making; whilst the classic texts of Austrian economics show why, in a world of uncertainty and perspective time, conscious decisions are better conceptualised as acts of plan-making rather than as choices (Shackle, 1972; Hicks, 1973). Essentially, this is because possible actions have first to be imagined, along with their possible consequences, and only then can a choice be made between alternative, subjectively imagined plans. That is, the set of options to be chosen between are not somehow objectively given. However the underlying set of possibilities from which that set emerges is objectively constrained. Ultimately, there are certain things which, in a given time and condition, it is not (yet) possible to imagine. Nor can the outcomes of options be *known*. They are only envisaged or imagined at the time choice is made. Moreover, the outcome of the path not chosen may never be known.

Now, it is an important truth (one stressed by the *Austrian* school of economics) that each individual makes decisions *subjectively*, based upon their own imagination of possibilities, and not as a kind of predictable or programmed calculating-machine, and that therefore both the range of plans considered to be chosen-between and the outcomes expected from each plan will differ between individuals, and thus so will their decisions and actions.

Moreover, it is also true (as stressed by the *evolutionary* school of economics) that each individual's conscious, rational decisions depend on what they learn from their own experience, and that, over time, the content of their subjectivity is formed by their interpretation of their experience, and is thus neither given nor constant, but formed and learned.

However, what results from this process of subjective decision making is not, for the most part, a chaos of completely idiosyncratic economic imaginings, or of continually-shifting economic relationships. There are two main reasons.

First, conscious exercise of individuals' economic imagination (of possible futures and possible outcomes for their plans) occurs only intermittently. Most of the time most individuals are, as it were, on autopilot, following habits and routines, repeating today actions that in the past led to a satisfactory and acceptable outcome, without generating unwanted explicit conflict or tension. Nor do individuals continually renegotiate all their 'social contracts' with those with whom they interact, as co-operators, subordinates or bosses in production, or as suppliers or buyers in markets.

Second, expectations of individuals are in large part formed by observation and imitation of the expectations, explicit or implicit, of others. Individuals thus absorb the conventional wisdom of the day, while this conventional wisdom is itself formed in large part by applying lessons learned from the past as guides to the future.

The interrelation between these two sides of the economic world is the proper focus of study for social economists: on the one hand between the worlds of the routine and of the consciously imagined set of plan options; and on the other between the worlds of stable expectations and of learning, innovation and originality of imagination.

The decision to purchase a building, or to initiate production of one, belongs for most economic actors in the realm of occasional self-consciously *planned* decision making. The partial exception to this are those firms for whom this is the content of their daily routines, for example, a volume speculative housebuilder. The question we must then ask concerns the respective roles of individual imagination and social conventional-wisdom in this property related decision.

Relation of a structure of provision to a structure of production

The need for the generation of stability in the context of individual decision making under uncertainty is the reason why the real world economy cannot operate as atomistic decision-taking individuals, but requires the institutions of organised markets and firms. The institutional approach to firms and markets is therefore in principle a general one, applicable to any part of the economy, though its actual content will be highly sector-specific.

Stability of institutions permits individuals to form both stable habits and stable expectations of the future. We shall call such a set of stable market and firm institutions a *social structure of provision* (SSP). More specifically, where a particular sub-set of stable markets, firms and individuals concerned with producing and transacting one product, such as new housing, is in question, we will speak of a social structure of *housing* provision, SSHP for short.

Some sectors submit more naturally than do others to vertical analysis (Ive and Gruneberg, 2000, ch. 1) in terms of systems of provision and production of a final product – in our case, the built environment or, within that, housing. Fine and Leopold (1993) offer a comparable study of two other, vertically defined systems – food and clothing. However, the durability of buildings, and the consequent dominance of the built stock by pre-existing rather than newly-produced buildings (and hence the total structure of building ownership and use) certainly raises an issue of importance here. It requires the concept of structure of provision to perform the theoretical work needed to form the bridge between the conceptual framework used for the study of the process of production of the new stock, the processes used at all relevant past times to produce stock still now in use, and the processes of exchange of ownership and use of the existing or second hand stock.

Whereas the key elements in a structure of provision are stable *social* relationships *between* types of organisation, the key elements in a structure of production are stable *technical* relations between workers, a knowledge-base and the physical objects and instruments of their work, and stable *social* vertical (authority) and horizontal (co-ordinated, collaborative) relationships *within* a labour process – between managers, technical specialists, production designers, production workers – some (sometimes, most) of which will be internalised within a single firm or organisation, and some of which will not be.

What correspondence will exist between a structure of provision (which locates a particular production process within the broader social context of the production, distribution, ownership and consumption of commodities, and the circulation of money and capital) and a structure of production? Occasionally something resembling experimental evidence becomes available to help answer this question. For example, a sudden shift or switch by a firm or set of firms from production of what is *technically* the same product within one system of provision to its production within another may be observed. In such cases, do we find that the firm(s) in question restructure their production process, or not?

Leopold and Bishop (1983) and Cullen (1983) report some interesting findings in relation to just this point for firms switching from 'public' to 'private' house building. They conclude that, even if product-design is held constant, production processes will in fact have to change quite radically – because of changes in methods of payment (cash-flow) or marketing, for example.

Certainly we take the view, on the basis of the admittedly less than comprehensive evidence known to us, from this and many other instances, that in the field of construction production the 'correspondence' is usually significant. That is, we think that, for the most part, we do not find a single dominant system of construction production used across-the-board, regardless of the social system of provision into which it is inserted. The social relations and techniques of production of 'social' housing are significantly different, we find, from those of production of 'speculative' housing – not, of course, completely so, but different enough for the difference to matter.

Concluding remarks

The 'structure' of both a structure of provision and a structure of production comes from the stability of institutionalised (individual) behaviour, which enables:

- conscious, would-be rational-maximising actors to assume that present institutional forms (types of market, firms and systems of relationships between firms) are a good guide to likely future forms;
- the assumption that the economic environment relevant to their decision but outside their control is stable and predictable, within limits, and that uncertainty is therefore parametric rather than structural. Parametric uncertainty refers to lack of knowledge of the values of the parameters of a decision-problem. Structural uncertainty is deeper, and suggests lack of information about the fundamental nature of the problem in terms of types of operative causes and types of possible effects or outcomes – the distinction is made by Langlois (1984), and relates to, but is not the same as, Knight's (1933) distinction between risk and uncertainty;
- everyday business life to comprise stable habits which are not, for the most part, experienced as producing disastrous or unsustainable results for their habitués, and stable expectations which their holders are not forced to abandon, even though they will, of course, only partially or approximately be fulfilled. Business managers are sometimes described as if their entire time at work consisted of making an endless series of decisions. In fact, even managers only take conscious decisions relatively rarely. Most of the time they too follow norms and routines and undertake actions capable of enabling activity to proceed without the need for conscious decision.

2
Social Structures Applied to the Provision of the Built Environment

Introduction

This chapter applies the concepts of social structures developed in Chapter 1. In particular the chapter investigates the implications of social structures in the production and provision of housing. We will divide economic actors *qua* decision makers into:

- the behaviourally most calculative or 'economic' (such as speculators),
- the lexical (such as consumers), and
- the most habitual (such as workers, subcontractors).

Previous institutional approaches to the construction/ development process

We are of course far from being the first to use some such concept of social structure or system of housing provision or production. Ball *et al.* (Ch. 5 'Property supply and institutional analysis'; 1998) contains a recent critical review of the relevant precedent literature.

Ball *et al.* begin by distinguishing organisations (firms, principal actors) from all other institutions. '(Non-organisation) institutions are the practices and networks that influence the ways in which those organisations operate and interrelate' (Rowlinson, 1997); and, in the metaphor of game-playing, organisations are the players while other institutions are the rules (North, 1990).

Using the term *institutions* broadly, to cover both organisations and institutions in North's sense, Ball *et al.* point out that the full list of institutions involved in the development process (their term for what we have called the construction process) is a very long one indeed, including, as well as the chief categories of actors – property developers, property owners/investors and lessees/tenants – others such as landowners, the utilities, land use planners, estate agents, banks, legal and financial advisors, architects, engineers of various specialist kinds, various types of surveyor (principally valuation, quantity and building surveyors), building material and component producers, construction managers and specialist- and sub-contractors (Ball *et al.*, 1998; pp. 110–11).

Following Gore and Nicholson (1991), Ball *et al.* identify (p. 113) several main approaches to representing the development process diagrammatically, in terms of flows and relationships, including:

- sequential, project-based models, describing the sequence of 'who does what' in a single project, from its initiation to its finish;
- structure/agency approaches (Healey, 1992; Healey and Barrett, 1990) using stock/flow multi-period models (for example, Barrett *et al.*, 1978). Each cycle of development activity in aggregate, commences from a given built stock and moves through a three-staged process of opportunity-identification, establishment of feasibility and development implementation, to yield (eventually) a flow of completed developments. This process transforms the stock and creates a new pattern of land use.
- 'structures of provision' models, in which diagrams are used to specify the key organisations involved in the structure, and the schematic relationships between them.

The approach we have used in the previous chapter is at first sight most nearly related to the last of these.

(A) structure of building provision (SoP) refers to the contemporary network of relationships associated with providing particular types of building. Those relationships are embodied within institutions ... and may take market or non-market forms. 'Provision' encompasses the whole gamut of development, construction, ownership and use ... Structure ... has a ... limited status, describing the main organisations and their relationships. It is used in the same way as in the phrase 'the industrial structure of a country' ... Prior theories are needed to examine the working of a structure of provision. Different types of (economic) theory could be used in conjunction with it ... Those theories could be used in various ways to explain the existence of an SoP, the factors generating its dynamics ... SoP can be said to constitute a methodology for examining institutions, rather than representing a theory ... in itself ... A structure of provision can be seen as a property market institution – the rules, practices and relationships that influence particular types of property development and use.

(Ball et al., 1998, p. 129)

However, we also share some of the chief aims and features of the structure/agency approach attributed to Healey, including the latter's empirical emphasis upon the local or the case study (see pp. 46–50). Healey's 'structures' comprise 'material resources', institutional rules and organising ideas. Her material resources comprise land rights, labour, finance, information – a list more or less identical to our 'overarching economic institutions'. The institutional rules are set by organisations or the political process and finally, the organising ideas are those which agents in the process acknowledge. These ideas inform the interests and strategies of actors as they define projects – a welcome emphasis on the role of economic imagination and consistent with the 'Austrian' approach to decision-making we endorsed above (Healey, 1992; pp. 34–8). Structures (in the above sense) constrain the behaviour of agents, whose behaviour is not 'determined' by the operation of competition, so that there is 'strategic' behaviour and real choice for these agents (again, we would agree).

Indeed, we would go so far as to suggest that the approach we develop in this chapter represents an attempt to fill some of the gaps in the existing agency/structure literature of the construction/development process, largely by attempting to combine it with various bodies of heterodox economic theory.

Social systems of organisation of the development or construction process (social systems of provision of built environment)

Throughout the following section, the reader is encouraged to make reference to Chapter 7 'Actors and roles' in the companion volume (Ive and Gruneberg, 2000). However, we have attempted to make the explanation and exemplification of our approach below sufficiently free standing, albeit concise, to be followable by readers of this volume alone.

The main actors in the production, ownership and consumption of buildings and the built environment include architects, engineers, contractors, speculative builders, property companies, public authorities and private owner occupiers (who can be either firms or households). These actors (mostly firms, but including all the main other types of economic actor, such as public authorities, households and unincorporated businesses) become involved in construction processes either as initiators (developers) or through direct or indirect contractual arrangements with developers.

The participants (actors) in the development process must undertake at least one of the roles of developer, designer, builder, owner and user, shown as column headings in Table 2.1. These roles are universal and apply to the processes of production of the built environment in all economies. However, the particular arrangement of firms (actors) on a project is specific in time and place. Moreover, the actors operate within a given legal and planning framework so that lawyers, town planners and even central government may also become involved in the process.

In Table 2.1 each letter (S, C and I) represents one social system of organisation of the development process – the speculative, contract and integrated systems, respectively. Each system is defined on the one hand by the way roles are either combined within one actor or fragmented with consequent contractual relationships between actors, and on the other hand by the social identity of the principal actors, the mode of governance of relationships between actors, and the locus of power within the system of relationships.

In System C, *the contracting system*, roles are fragmented between three chief actors – the developer or client, the architectural or

Table 2.1 The roles of participants in the process of providing the built environment

Transactors	Developer	Designer	Builder	Owners	Users
	Roles				
Speculative builders	S	S	S		
Contractors			C		
Architectural firms		C			
Engineering design firms		C			
Owner occupiers	C			C, S	C, S
Public authorities	C, I	I	I	C, I	C, I
Property companies	C			C	
Tenants					C, I

Note: S = speculative system; C = contract system; I = integrated system.

engineering design firm and the contractor. The locus of power is likely to rest (at least, before contracts are signed) with the client, that is, the developer, though part of this power may be delegated to agents of this principal (particularly, to firms of designers). The developer's power will be the greater, the more the information about the construction process (knowledge of the identity and costs of the most efficient producers and methods of production, for example) held by that party. Some actors holding this role may utilise that power in different ways to those of others (for example, public authorities cf. property companies). The location of power post-contract, by contrast, may rest mainly with the actor holding the role of builder (the contractor), unless the developers are repeat clients, who can therefore hold out prospects of access to future projects to their contractors.

Contracts to design and to build are signed before production begins and before actual costs are known. The risks and uncertainty connected with construction costs and values (in the market or value-in-use) can be substantial. Yet sales precede production. Some part of this risk is therefore borne by the client, who pre-commits to pay (at least) a fixed price. This is in stark contrast to much of manufacturing industry, where goods are offered for sale only after production has taken place, and where risk is therefore borne by the producer. The importance of contracting and the contracting system is one of the most significant features of the construction sector.

In System I, *the integrated system*, roles are of course integrated into the hands of one actor, though they may then become divided into the special responsibilities of departments of that actor. The overall locus of power is clear, though there may be differences in the balance of power between departmental interests within that dominant actor. One classic example of the integrated system in Britain was to be found in many local authorities from the 1920s to the 1970s, where user-departments (such as Housing or Education) commissioned and took ownership of projects provided by developer-departments (Estates, Finance, Development) using the services of in-house design and construction departments (Development Services, Architects, Direct Labour Organisations), under a wide variety of actual intra-authority organisational arrangements, though cost risk in any case was borne by the public authority as such (value-in-use risk being shared between owner-department and users/tenants).

In System S, *the speculative system*, roles are partially integrated. But the dominant feature is the exclusion of the final owner from the role of initiator or client (developer). The locus of power is with the developer, who will also perhaps (though not necessarily) hold the roles of designer and builder. Production precedes sale, and market value and cost risks (and rewards) are borne by the speculative developer. Note that even if a speculative builder places contracts for design and construction with separate firms of architects and building contractors, it is possible to still consider that instance of the building process to be organised under the speculative system. There are two arguments for this treatment. First, that it is right to distinguish cases where the developer intends to sell on completion (and is thus a speculative developer) from ones where the developer intends to own the building. Second, that in practice in the modern UK most speculative house developers (though not in fact most speculative developers of offices or other kinds of building) are speculative builder/developers – that is, do in fact retain effective control and management of design and construction, and are correctly treated as a species of construction firm, and as part of the construction industry, unlike other kinds of speculative property developer. Thus any contracts they place we regard as 'akin' to the subcontracts that a main contractor in the contracting system places, rather than as indications that fundamental responsibility for basic roles in the process is fragmented.

All three basic systems (contracting, integrated and speculative) are in fact best thought of as families of similar variants, as genuses containing several species. Thus such arrangements as *management contracting* or *design-and-build contracting* we would initially regard as species of the contracting system genus. However, we need to be alert to the possibility of the emergence of species not belonging to any of our pre-existing categories of genus. Thus, for example, in the case of design-and-build contracting, a key test would be the extent to which full responsibility for design and construction is actually given by such contracts to an integrated D&B firm. In principle, if it were, then there would be a case for recognising a new genus (just as there might be for DBFO – design/build/finance/operate – contracts, characteristic of the Private Finance Initiative).

There is one further feature which helps distinguish and characterises the contracting system – its organisation of the process in terms of individual projects, rather than in terms of programmes or sets of projects. Both the integrated and speculative systems, in contrast, are structured so as to deliver programmes, not single projects – and are most efficient as forms of organisation where developers do have such programmes.

Because of each project's size, any one project often becomes a major source of income for the builder, especially in the speculative system, though also in the contracting system. Each project can form a relatively large proportion of a contractor or speculative developer firm's turnover, exposing the firm to significant losses in the event of a project's failure, discontinuity of work at the end of a project and uncertainty of future workloads. Consequently, each project is both an opportunity for a firm to develop and grow as well as a threat to the firm's survival.

Two social systems of new housing provision: speculative and contracting

Because the concept of a social system of organisation of the construction process (SSOCP) is both complex and novel, we propose to illustrate it at length by application to one important and complex branch of construction – new housing. In this context, we propose to use the term: social system of housing provision (SSHP). Of course, much, but not all, of the discussion applies equally to other structures of provision within construction. Thus, once this particular application of the idea has been studied, it should be possible to attempt parallel applications, developing equivalent special cases of the broader concept of a SSOCP.

In contemporary Britain, we identify two main SSHPs. In one current SSHP, which for want of an established name we propose to call the 'contracting SSHP', one type of client (social housing associations) and one public authority responsible for the regulation and funding of these clients (the Housing Corporation), together have devised a particular set of 'rules of the game' for the formation and placing of contracts, institutionalised a market, encouraged the formation of a specialist population of suppliers (architects and contractors), and developed specialised market-serving institutions. Housing Associations use certain characteristic types (design-and-build) and legal forms of contract and contract provisions (payment arrangements, etc.).

In another current SSHP, which in shorthand we call the 'speculative SSHP', it is the speculative builder/developers, more particularly, the 40 or less firms known as 'volume' housebuilders, each of whom builds 500 or more dwellings per year, who together account for more than half of all dwellings built in this SSHP[1], who are the key actors. These volume housebuilders again set the rules of the game, sit (spider-like) at the centre of the web of transactions that constitute the framework of the SSHP, and benefit from specialised market-serving (making) institutions, developed to facilitate their characteristic transactions and to supply them with information.

The 'speculative' social structure of housing provision (SSHP)

Habit and expectation in the pre-1990 private housing market

One stable expectation in the UK (and especially London) private housing market of the 1950s through to about 1990 was that house prices would rise over any substantial period of time faster than the RPI, faster than the effective rate of interest available on household savings in acquisition of non-equity financial assets, and possibly even faster than money incomes of households.

This stable expectation became associated by the 1970s and 1980s with certain stable assumptions and habits of households: the habit of borrowing as much house purchase credit as financial institutions were prepared to lend; the habit of buying the most valuable housing asset that could be afforded, in terms of a budget constraint on debt repayments at the peak of their expected negative cash flow profile; the habit of treating capital gains and dwelling asset values as readily liquidisable assets, either through sale or by borrowing against house equity collateral; the habit of transacting exchange into ownership of more valuable dwellings as income was anticipated to rise, over the earlier phases of a predictable household life cycle; but also the habit not (to the same extent) later, symmetrically, to 'trade down', because debt repayment would be complete by that point, and income expected to be sufficient to cover other expenses; thus, deriving from the last point, the habit of children of assuming they would inherit substantial net worth of dwelling assets from their parents, and conducting their own behaviour under this assumption.

All of this was prevented from inducing truly frenetic frequencies and volumes of transactions only by two things, the one a habit and the other a transaction cost. The habit or routine in question concerns the tendency to quotidian repetition in everyday life, based in part on inertia and in part on the desire and need for some stable anchors in a changing world. It can be summed up in the idea that 'a house is a home'. Thus, once a person or household have moved house, they will settle there for a while, and behave as if it were their permanent home, suspending their asset-owner's calculus of advantage from further transaction for a while. The transaction cost in question is the literal transaction cost of fees to agents involved in market-searching and transfer of property rights using estate agents and solicitors, respect-ively, plus costs of physical transportation and storage of household goods, plus time and stress costs incurred directly by the transacting household.

However, despite this inertia and transaction cost, as a result of the prevalence of the above assumptions and habits about asset-acquisition behaviour amongst owner-occupiers of housing, other actors in the SSHP had their own less conscious habits and expectations affected. These other actors were also in a position where even their most con-sciously researched, explicit decisions could be (and were) based upon assumptions that this kind of owner-occupier behaviour as purchasers, sellers and borrowers would continue.

The housing market was known to be cyclical, of course, both in its price levels and volume of transactions, but cycle downswings were believed to be manageable and survivable by firms in their impacts by following the practices embedded institutionally in the key types of firms active in the market, practices learnt from the past and validated by survival of their practitioners. These practices included land banking and contra-cyclical net land acquisition, heuristics for 'safe' gearing and interest-cover ratios, and flexible rates of development of large sites.

Lexicographic decision-making by owner-occupier purchasers and product market segmentation

One of the actions of a household most likely to be the subject of lengthy, researched decision making of the kind implied by orthodox economic choice theory is the selection of one dwelling from amongst the population available for purchase. Earl (1983, 1986) provides an account of just this case from a lexicographic/behavioural economics approach. Behavioural economics (following Simon, various dates) stresses the assumption of bounded rationality. The theory of lexico-graphic decision making adds to this assumption a further assumption that buyers perceive products as offering combinations of potentially desirable features. Faced with x products each with y features, 'rational' consumer choice would require decisions to be made by simultaneous comparison in a y-dimensional decision space. Unable to do this, but faced with an important decision worth spending some time over, buyers apply hierarchical and sequential rank ordering of features and use exclusion-heuristics (rules of thumb) at each successive stage to simplify their choice problem to manageable proportions (e.g. 'won't look at any house with an asking price over £100,000'; 'look only within the M25'; 'reject if no garden'; etc.).

It then becomes worth a producer's while to investigate whether or not many consumers use similar feature-cognition frames, rank order-ings and exclusion heuristics. If they are found to do so, then that group of consumers can be efficiently targeted by the producer, who can design, and supply information about, their product so as to increase its chances of selection disproportionately to the cost involved in doing so.

During the 1970s and 1980s the UK owner-occupied housing produc-tion system was in a sense reaching maturity, in that the rapid growth in aggregate volume characteristic of the earlier post-war period was slowing down. This resulted in stagnant or diminishing sales for stan-dard products of types that had dominated the 1950s and 1960s. At the same time, however, it increased the pressure and incentive for firms to extend their product range. In this context product range means number of combinations of site- and dwelling design-characteristics.

Now, consumer choice of housing is made essentially by applying lexicological selection to a range of dwellings of which new dwellings are only a small part. The seller of new dwellings, therefore, first and foremost faces the task of giving them perceived and actual features, in comparison with the features of existing housing, that will help them survive such search processes for sufficient consumers.

At first sight, it appears that they have been remarkably unsuccessful in doing so. That is, not only do new dwellings (either on average, or for 'type' denominated as terraced/semi/detached house/flat) not command a price premium relative to existing alternatives, there has sometimes even been a price discount (DETR/DOE, *Housing and Construction Statistics* (hereafter, *HCS*), annual). This was certainly common in the period dominated by green-field, ex-urban develop-ment. Were such new dwellings judged locationally less desirable (less proximate to important established centres of work, shopping, leisure, etc.) than most existing ones? Only if so can this price gap be explained in a manner consistent with normal or super-normal profit for volume developers of new green-field housing (which requires that this hypothesised relative unattractiveness of site locations is fully reflected in lower land acquisition prices, relative to site values in exist-ing urban areas). Green-belt containment policy in the South East probably heightened this price gap, by forcing new development beyond the protected ring, rather than allowing accretion to edges of established suburban locations.

In the 1980s, much more than before, developers succeeded in developing site-cum-dwelling design combinations capable of commanding price premiums over local existing dwellings in the same first-order type category. In other words, they successfully launched 'new' product ranges onto the local markets (especially and initially in London), widely perceived as 'superior products' by enough purchasers.

Economic speculators, land markets, land acquisition decisions and long-term expectations under uncertainty

Let us assume that land prices and transaction volumes for land of a given type and location are set as follows: for land bid prices are formed by bidders based on guesstimates of future returns. These valuations correspond to Keynes' long-term expectations of investment returns. Given radical uncertainty, they are best modelled in terms such as Shackle's potential surprise (Shackle, 1961, 1968), together with an account of how waves of market sentiment can arise irrationally. 'Irrational' (given radical uncertainty and bounded rationality, valuations can be of no other kind) valuations are affected by market sentiment and the psychology of 'following the mass of others' (Keynes, 1936) but also by institutionally-structured information and expectations endogenously-formed within the institutions of the system of provision. The conventional account errs in assuming that rational calculation of probabilities can produce a single value of maximum bid per bidder. 'Austrian' subjectivity helps, by allowing different bidders to have widely different subjective expectations and therefore valuations, but is not enough. Sellers' reserve prices (absolute rent) mainly affect volume of transactions in land, and set a price floor; but actual transaction prices are set by imperfect bidder competition in conditions of imperfect information and uncertainty.

Confidence to make land (and, for that matter, house) purchases even under uncertainty is improved by the existence of well developed, regularised markets, to give some reversibility of decisions. But note that reversibility is much less secure than with financial assets (liquidity of the asset is less).

Here, developers' investment decisions to acquire land represent just one type or case of investment decision within a SSHP. A research programme adopting this approach would seek to study the effect of structural stability and institutional context on all those decisions, via formation of expectations and behavioural decision heuristics.

Housing markets, housing acquisition/disposal decisions and long-term expectations under uncertainty

In considering long-term asset-acquisition decisions by housing purchasers in a context of housing markets, we may have a particularly apposite field for application of Keynes' model of stock market behaviour (Keynes, 1936). In the financial asset market context, Keynes has since been criticised for underplaying the role of futures markets in dampening 'sentiment'-induced wild swings in stock valuations. However, in housing, no such futures markets exist. Thus there are less grounds for departing from Keynes' characterisation of how values are set in a market in financial assets: the best strategy and likely psychology of each transactor will be to 'follow the herd', in the absence of certainty and adequate information on which to form, and have confidence in, their own independent opinion. 'If the market is rising, because there is a flood of buying, the market is liable to be right, or rather, I have no better basis than assuming others know what they are doing on which to base my own decisions, and so it is best for me too to buy.'

Housing market cyclical upswings, once begun, continue for a while. Once asset prices start rising, the upward trend may be immune to various pieces of new information. Minor bouts of selling and short term price falls will be ignored by 'the herd', and market sentiment will remain optimistic. However, once some new information (unforeseen events) causes a sufficient surprise to force many or important actors to reform their basic long-term expectations, and these actors then start selling and cease buying as a result, their new pessimism, validated and reinforced by the resulting initial fall in prices, spreads to become the new shared sentiment of the market, and prices continue to fall, on a sustained wave of selling. Once prices begin to fall then, given the lack of a futures market, many buyers decide to defer purchase, in expectation of further falls; thus initially demand *falls* (leftward shift in demand curve) as a result of the price reduction.

Prices do not fall automatically when the number of *ex ante* sellers exceeds the number of buyers at *ex ante* price, because the housing market does not clear in each time period; rather, they fall when and if, as a result of unsatisfied intentions to sell at *ex ante* price, sellers change their instructions to their 'brokers' (estate agents) about reserve price or asking price. Otherwise, it is the number of transactions that falls, with consequent loss of potential liquidity for owners of housing assets.

One question is, how best to think of what is meant by 'sellers' or 'intention to sell', in a context in which virtually all sellers intend to re-purchase another housing asset? Though sale will be followed by a purchase, the point is that the purchase is conditional on prior successful sale (we assume that bridging loans are regarded as prohibitively expensive and risky where sale and sale price is regarded as uncertain). Say's Law once again fails to hold; supply does not create its own demand. This would apply *a fortiori* if most sellers planned to trade-down, but this is not necessary to yield the result.

It should be noted that one, apparently paradoxical, aspect of this analysis is that long-term expectations are treated as rather stable and consensual at the same time that emphasis is placed on the mutually inconsistent and unstable nature of speculators' short-run expectations. We do not however regard this as anything other than an attempt to capture the real case of things.

Intermediaries, speculators and information asymmetry in housing production markets

Housing production markets contain large sets of 'professional transactors'. In the 'speculative' SSHP the chief of these are speculative developers and their agents, who make their profits out of such transactions, and depend largely upon being able to 'out-perform the market', and thereby to increase the net worth of their land and housing assets and money capital faster than the general rate of house price increase; or at least, faster than the cost of their borrowed capital. Such speculators have the incentive, means and opportunity to research their decisions to a much greater extent than occasional or amateur transactors, and have access to private sources of information. Thus we have a classic case of information impactedness and asymmetry (Williamson, 1975). Sometimes these professional speculators are also active in some other role in the production process (as infrastructure developers, promoters, producers) but sometimes they appear only as traders. The latter is most common with existing housing with redevelopment potential, and with vacant or change-of-use land. Most newly built dwellings are, in fact, sold direct by speculative builder-developers to owner-occupiers. The following passage will therefore cover both speculative traders and speculative developers.

The importance and prevalence of information asymmetry in housing markets can be illustrated with some examples. There do exist 'open' markets in housing land. However, most land acquired for housing development is transacted 'privately', before its potential availability (willingness of former owner to sell) or developability (probable treatment by planning authorities) becomes widely known (i.e. becomes general knowledge relatively costlessly available to any searcher), and not through public auction. Estate agents and solicitors benefit considerably from possession of 'inside' knowledge. Indeed if information was not so impacted, the housing market could hardly have supported such vast numbers of 'information professionals' (estate agents and solicitors). Knowledge of identity of the owner of a particular tract of land is not public knowledge. Even if the tract is covered by the Land Registry, access to that data is restricted to trans-action-professionals. Shifts in the balance of probability of planning authority treatment of development applications in an area are often first learnt-of by market 'insiders', even before they may filter through to all existing land owners – and the case is likewise for the knowledge of interest of a potential final buyer in a site. Proposals to create public-access data bases with a maximum amount of information on past transactions and present detailed actual and potential supply and demand have been vigorously and successfully resisted by the transac-tion-professionals, both agents and principals. Former owner-occupiers of sites which become development opportunities rarely sell direct to the new final owner. Nor do these former owner occupiers, such as farmers or housing owner-occupiers, often participate as equity partners in development. Now, in many cases this could be explicable in perfect market terms. It could be a matter of liquidity preferences and risk aversion. However, its very prevalence casts doubt on this explanation, and suggests that informational barriers create the 'space' between seller and ultimate buyer for intermediary traders to occupy.

As well as information asymmetry, however, we must also stress a more 'Austrian' explanation of the same phenomena, viz. difference in subjective present valuations of expected future sale prices of assets. In this model, the same information is available to all, but it is 'inter-preted' differently, as a result of different assessments of the uncer-tainties involved. Any given market valuation of an asset sustains both 'bulls' (those who expect its future price to be higher, and are therefore willing to buy at the current price) and 'bears' (who expect price to fall, and are therefore willing to sell), and moreover in roughly equal numbers. For current price to be stable, demand to buy (from bulls) must be matched by willingness to sell (from bears) (Shackle, 1972).

Divergent expectations between classes of actor in SSHPs

Veblen (*Instinct of Workmanship*, 1964; *Engineers and the Price System*, 1921) was perhaps the first to develop a whole economics grounded in the assumption that different economic classes diverge in their modes of calculation and of formation of expectations. He distinguished on the one hand between entrepreneurs and others; and also between owners of finance and managers of businesses. SSHPs offer promising terrain to develop this. Also, it would invalidate a 'rational expectations economics' approach to the subject if it were to be demonstrated that in this field at least consumers, production firms and financiers hold divergent expectations, based either on divergent 'underlying theoretical models of how the economy works' or on different processes of expectation-formation and adaptation.

Let us consider the case of housing asset 'positions' of two classes of actor, owner-occupiers and housing developers. The Austrian approach (Bon, 1989) is for each actor to be assumed to review the structure of their asset portfolio continuously, in the light of new information, as they interpret and evaluate it. Thus transactions require that some actors are seeking to increase their 'positions' in (planned holdings of) a certain type of asset whilst, at the same time, others are seeking to reduce their positions, and shift towards other assets, including cash. If there is a net imbalance between the volumes of the two sets of plan-adjustments, then there will be a change in the market price of the asset. In the light of this price change, planned asset positions will be further reconsidered; and so on.

Now, is it the case that, towards the end of a boom in housing prices and production, developers and other speculators decide to reduce their 'positions' at a time when owner-occupiers collectively are still trying to increase theirs? For, if there are divergent 'class' expectations, we would hypothesise some such result. For a developer, 'reducing their position' must be interpreted to mean, selling a larger value of assets in period *t* than they acquired during that period. What determines the value of housing assets acquired by a developer in a period? Since we draw a category-distinction between land assets and housing-assets (we regard such developers as making choices about division of their asset holdings between land, housing, cash, and other financial assets – land and housing are different assets because they can be sold in different markets, and different factors will affect their realisability into cash and their price), it must be thought of as either the estimated market value on completion of the dwellings started in that period, or the increase in the cumulative value added, measured at cost, of dwellings under construction. Austrian school economists would of course use the former. Let us follow them in this. We then have the question for a potential piece of research, aimed at acquiring information on developers' subjective money valuations of 'starts' and comparing these with both expected and actual values of developers' sales of dwellings; and then comparing this net balance with the net change in planned position of the class of owner-occupiers.

The underlying hypothesis is that developers and other speculators have access to information that enables them to try to anticipate turning-points in the market, whereas owner-occupiers work with lagged information about *ex post* price index changes. This would imply either the existence of 'lead' indicators more closely and more quickly monitored by developers than others, or developers' possession of predictive models. However, there is anecdotal evidence (for example from interviews with developers) to the effect that in fact they too only respond to, rather than try to anticipate, market turning points. Such a line of inquiry may therefore be a cul-de-sac.

Nevertheless, if we take out the majority of transactors in any period who are neither significantly increasing nor reducing their positions, but merely relocating, we are left able to focus on those who are adjusting their positions. Speculative readjustment by transaction-professionals may not show professional market sentiment at any one time to be equally divided between bulls and bears (the normal assumption for price setting in a speculative market for durable assets) – instead the imbalance could be made up by non-speculatively motivated net transactors.[2]

Role of firms as institutions in SSHPs

Transaction cost economics tries to explain the location, in any particular case, of the firm/external-transaction boundary. At one extreme, we would have a completely vertically integrated SSHP in which each firm would combine all roles short of eventual buyer of a commodity flow, or tenant. At the other, there would be an atomistic fragmentation in which firms would tend to be little more than one-person businesses, each specialising in one sub-role, engaged in a massive net of inter-firm transactions to produce the commodity.

The speculative system of housing production in the UK is now much more highly integrated than most systems of construction, in terms of concentration of authority and power in a single firm, the builder/developer. However, it is also characterised by subcontracting, and marketisation of the employment relation (temporary, 'spot' contracting for labour).

The firm can be the key institution for reconciling decision making with radical uncertainty. The firm can do this by creating (investing in) internal flexible resources, capable of responding to changed expectations. The firm in other words constitutes a pool of strategically internally reallocatable resources. Uncertainty makes the transaction costs involved in using markets very high. Uncertainty cannot be covered by contingent contracts (Kay, 1984; Langlois, 1984). 'The function of the firm is, therefore, not simply to minimise transaction costs, but to provide an institutional framework within which, to some extent, the very calculus of cost is superseded', (Hodgson, 1988, p. 207) – that is, superseded by conventions, rules and routines not defined by market price information. 'Its relative internal stability means that the firm can carry unquantifiable risks which would be eschewed in the volatility of the market', (Hodgson, 1988, p. 213), especially investment in R&D.

The problem with all of this, as applied to firms in SSHPs, is that here we find firms apparently intent, if the above line of reasoning is correct, on organisational euthanasia. If firms deal with uncertainty only by externalisation and continuous short term recontracting, eventually they lose their *raison d'être*, in the sense of their superiority over atomistic market-only organisation. Firms which cannot insulate their employees to a considerable extent from the vagaries of the fluctuating market offer little to them, and will find it hard to develop the advantages following from durable, trust-relationships. (If economies of scale of production costs are very great, the force of this point is weakened, because any large firm is then able to offer higher wages to employees than would be available in self-employment as an independent small producer.)

Ive (1990) offers an application to construction firms of a model of the process of formation of a firm's objectives, and the evolution within it of dominant conventions and modes of calculation, given that each firm is a structured institution (or, 'dispersed social agency', in Thompson's terms) combining powerful individuals and groups with diverse objectives and modes of calculation. Following Thompson (1986), in Ive (1990), it is taken as axiomatic that all systems of calculation and measurement, including accounting and investment appraisal, used in a firm are conventional, discretionary, not 'natural', and are influenced by power and ideological dominance in this dispersed social agency. Beer (1972) is also highly relevant, making the same point within a 'systems' framework of thinking about the firm. Ive (1990) suggests that firms are best modelled institutionally in terms of balance of power between functional departments (production; R&D; marketing and purchasing; finance and treasury) each constituting an 'interest', from which strategic objectives can be imputed. This balance is seen in turn to depend partly on the external environment, including systems of provision of commodities, in which the firm finds itself, and partly on internal traditions, norms, 'culture', etc., of the firm itself.

How many SSHPs are there/have there been?

How do we distinguish between adaptation of an existing SSHP and emergence of a new one? Hodgson (1988) draws on systems theory for its idea of the 'law of sufficient variety'. He terms the result of his analysis a demonstration of 'the impurity principle'. Systems containing insufficient variety will eventually meet contingencies with which they cannot cope. This will produce chaotic system-breakdown, and require more-or-less deliberate invention of a new system.

On the other hand, 'impure' systems, comprised of many parallel systems, co-existing with the currently dominant system, have Darwinian virtues of evolutionary capability (Darwin, 1929 edn, pp. 82–100). Environmental changes outside the range to which the dominant system can adjust, can be met by increased resort to one or more of the formerly subordinate systems. This does not imply necessary regression to a 'former' system. Though it is the case that many subordinate systems will be survivals of formerly dominant ones, even these have capacity to adjust and adapt within a range, and can be expected to do so under the new contingent environment. Moreover, some subordinate systems may have a quite different history and origin (attempted innovations which failed to achieve dominance but yet survived).

Methodology of a systems approach

Methodologically, our decision about system-splitting and sub-division will be crucial. If we think systems can adapt and 'split' into variant sub-systems, then much of the power of the systems-approach may be lost. What is crucial is whether or not the distinction between limited adaptation, within a system, and system-replacement becomes blurred.

Whether we 'see' in certain changes a new system, or only an adaptation of a continuing system, will depend in part upon the historical time-frame and perspective we adopt. In one perspective, the 'speculative' and 'contract' systems have each existed in parallel now for over a century. The key historical event is seen as the institutionalisation of the two systems in conscious and regulated distinction from one another in the first half of the nineteenth century (Bowley, 1966). Since then each system has changed and adapted somewhat, but has retained crucial defining features. Moreover, each system is seen as characteristic of the whole range of built environment provision, and its persistence is seen to represent the 'special' conservatism and inertia of institutions in the construction process.

In another perspective, the importance today of that early nineteenth century bout of institution-formation is much less. As the wider economy has developed through successive broad structures of accumulation, so has the construction world. Nominally (i.e. in terms of the names and language used), the 'contracting system' of the end of the twentieth century may appear to be in direct line of descent from, and essentially only a variant of, the 'classical' contracting system of the nineteenth century, but the reality is quite different. The modern construction contractor does not work under the 'tutelage' and paternal guidance of modern-day Brunels or Stephensons. Architects no longer command the industry. The dominant clients are no longer the court, church and aristocracy, nor individual entrepreneurs. A modern Housing Association shares virtually nothing with its nominal predecessors, the nineteenth century Labourers' Model Dwellings Associations. What is needed, according to this perspective, is a willingness to recognise and identify a succession of historical periods, each with its own social structures. Periodisation has its problems (lack of agreement on the criteria for identifying and dating periods; lack of 'correspondence' between periods that 'fit' at one level of abstraction, such as the world economy, and those that 'fit' at another, such as the UK construction process), but is nonetheless the essential step.

When these two perspectives are posed starkly as either/or exclusive alternatives, the choice offered is invidious. The more sensible course is surely to insist that it is rather a matter of both/and, not either/or. Each 'new' period correctly has its modernity proclaimed, but this does not mean that the preceding institutions are obliterated, that the earlier history is no longer of any consequence.

The position taken here is that it is better, on balance, to be prepared to recognise new systems as new and not as splits from or variants on the same 'genetic' root, but recognise also that usually they will be built-up by salvaging elements from old systems. However, the point is that elements are defined relationally. Therefore, once contained in a different system, these elements become different (acquire new meanings, determinants/constraints, expectations, etc.). There is an analogy with Schumpeter's (1934) definition of an innovation as a 'new combination'.

On the other hand, system-description model diagrams (Ball *et al*, 1998; Gore and Nicholson, 1991) which represent one of the main ways to date of analysing systems of housing provision, are certainly vulnerable to being read as suggesting that a new 'system' emerges whenever there is one significant change in the set of relations between actors. A methodological problem would arise if it is a matter for *a priori* untested judgement by the analyst as to whether a given change is sufficiently major as to institute a new system.

Here we propose that it may prove useful to distinguish different *degrees* of change in the major assumptions embedded in the set of basic long-run expectations of the main actors in the system, and the structural features of the *spatiality* of a system of provision. Some such changes can be considered changes within a system. If the system proves able to adapt to and absorb them into the same basic set of relations between actors, then the same system can be said to persist (in variant form).

The most important need, if this 'structure/agency' approach is to develop to fulfil its aims, is to develop a measure of the total net of relationships between actors, weighted by importance of the relation, so as to be able to distinguish system-replacing from other changes in the set of relations between the actors constituting a system.

An application of 'systems of provision' approach to metropolitan SSHPs: the case of London and South East England

This application is at the level of a region of the UK. This is an attempt to enable us to explore the spatial as well as the social structure, and the significance of the spatial in defining an SSHP. It is a response to the impossibility of generalising about spatial structure at the national level, at least in this case.

Notation:
(A) = a class of actor

$\Sigma(A)$ = the set of relations between actors constituting a system
(S) = structural features of the spatiality of the process of provision
(LE) = major assumptions embedded in a stable set of basic long-run expectations about the environment of the system, and its performance
(FR) = regime of financial regulation (financial rather than other regulation, simply because housing production always involves financial investments and borrowing, and is therefore particularly effectively regulated in this way. This is not meant to imply that there are not other important dimensions to regulation. Spatial regulation is subsumed under S, however.)

Post-war period
Thus, in the 1950s–1970s ('post-war') period, we could identify:

System (1): the 'speculative' SSHP in the 'post-war' period.
A dominant SSHP, accounting for most (and a secularly growing share) of production, comprising:

Actors (A)
(A) owner-occupiers (behaving as a mixture of consumers and asset-acquirers);
(A) 'volume housebuilders' as described by Ball (1983);
(A) direct production subcontracted to LOSCs, with 'lump' payment systems;
(A) building societies (BSs) making variable-interest long-term mortgage loans to owner-occupiers (O-Os);

Spatial structure (S)
(S) green-field development within a regime of urban containment and permanent land shortage;
(S) dispersal of population at regional level;

Long-run expectations (adaptive; 'conventional-wisdom') (LE)
(LE) modest, short cycles of output and price about a slowly rising trend level of output and strongly upward price trend;
(LE) periodic credit rationing imposed by building societies, and general Building Society 'management/regulation of an orderly market', putting limits on 'potential surprise' expectations;

Financial regulation. (FR) total purchaser expenditure regulated by Building Societies and Bank of England

System (2): the 'contract' SSHP in the post-war period

A 'survivor' SSHP, briefly formerly dominant (from 1945 to the mid-1950s) comprising:

Actors

(A) local authority (LA) tenants of new dwellings, drawn increasingly from 'homeless' need-category (cf. slum clearance and waiting-list), and therefore increasingly politically powerless and unorganised;

(A) matrix development bureaucracies, comprising weak (in this context) client (Housing) departments and strong functional delivery departments (Technical Services, Development Services);

(A) mainly LA in-house production of design, contract documentation and contractor-supervision;

(A) major and medium sized building contractors, producing under Bowley's 'traditional' contracting system of relationships between contractors, designers and clients (Bowley, 1966);

(A) rapidly increasing involvement of subcontractors, both supply and fix and LOSC, chosen by and under authority of main contractors, but with some 'nominated' subcontracting also;

(A) banks engaged in general lending (mainly purchase of medium and long-term bonds) to central government and to LAs, some of which was then used by them as development finance (NB: banks' loans not tied or linked in any way to projects or to dwelling stock as collateral, but secured on general tax revenues);

Spatial structure

(S) increasingly smaller project sizes, on in-fill sites;

(S) regional production increasingly concentrated into Inner London;

(S) breakdown of region-wide or supra-LA mechanisms for housing production within this system, such as GLC/new towns/expanded towns;

Long-run expectations

(LE) initially, cycles of output about a stable trend but then, at any rate after 1976, and possibly as early as 1969–70, expectation of long-term decline in output;

(LE) endemic lack of capacity relative to demand;

Financial regulation. (FR) LAs regulated in total expenditure and in resource-allocation to projects by a triple-regime of central government controls: total permission-to-invest controls; subsidy controls, both over subsidies to LAs (central government transfers to Housing Revenue Accounts) and to tenants (housing benefit/rebated rents); and project-approval controls (mainly, Housing Cost Yardsticks).

The neo-conservative or current period

Between 1980 and the late 1990s the dominant 'speculative' SSHP system ((1), above) showed considerable adaptation:

a) Within this system, its major adaptation to external contingency was the 1980s adjustment to financial deregulation, with consequent opening of a formerly closed system boundary, between the dominant SSHP, including its own financial institutions, and the wider financial system;

b) Another adaptation involved increased product segmentation, and new spatial diversity, in response to slow down and stagnation of total output trend, and intensification of 'green field' land shortage.

System (1) therefore displayed increasing internal diversity, in that new forms of product and spatiality of production did not immediately replace all older forms, which continued, but now in relative decline.

However, during the same period of the 1980s, we argue that, whilst System (1) adapted to new contingencies, System (2) was simply destroyed. A (Thatcher government) contingency arose with which it could not cope. No adaptation would have been sufficient to ensure survival.

System 3: the 'contract' SSHP in the neo-conservative period

Also, mainly in the same decade, though with origins in the 1970s, a new System (3) emerged. Also a 'contract' system (as opposed to speculative), nevertheless in our view it represented a new system, rather than an adaptation of System 2. It was, in effect, one of a limited number of possible forms which a 'contract' system could take (could have taken) within the constraints posed by the overall social, economic and political structure of the period.

System 3's key new institutions, in the 1980s, were:
(A) not-for-profit housing landlords in the form of Housing Associations (HAs), either originating in user-organisations or charities to serve particular user-groups, or as autonomous manager-controlled provision organisations;
(A) professional service firms selling design, surveying, project management services to HAs;
(A) contractors displaced from System (2) plus smaller contractors, working increasingly under new forms of the contracting system, such as design-and-build;

(A) new tenants selected either from defined need-groups or 'nominated' by LAs or partially on ability to pay break-even rents;
(A) banks, etc. supplying development finance mediated via Housing Corporation;
(S) as for System (2);
(LE) of secular reduction in amount of subsidy per dwelling, tight cash limits on overall subsidy, switch towards public/private financial partnerships, and restriction on growth in total volume of production;
(FR) via the Housing Corporation.

One hypothesis is that in the early 1990s System (1) began to show signs of breakdown, as it encountered conditions beyond the range to which it could adapt. Amongst the evidence for breakdown would be: new (LE) for house prices incompatible with some basic assumptions ('certainties') embedded into the old system; negative equity, defaults and repossessions; loss of realizability of housing-assets; increased insurance costs of borrowing; shift in perceived relative returns, risks and costs of mortgage-financed owner-occupied and other forms of tenure/finance; conversion of some former speculative builder/developers into non-speculative suppliers to System (3); etc.

Now, this does not mean that the hypothesis foresees 'the end of owner-occupation' or the 'end of private house building', but only the end of the dominance of a particular combination of actors and roles. If this hypothesis is correct, then new institutions will be needed, to permit formation of a new system, comprising a new structure of relationships.

Alternatively, we have the hypothesis that System (1) is still within, though perhaps near the edge of, its range-of-adaptation.

All the above argument has benefited enormously from a critical reading of the works on crisis, structure and emergence of new structure in the recent American Social Structures of Accumulation literature (Kotz *et al.*, 1994; Bowles and Edwards, 1993; Gordon *et al.*, 1994). In that literature, however, perhaps the weakest link is in the assumptions made about how a new structure of institutions emerges. The danger is a functionalist logic in which 'the system' faced with crisis somehow diagnoses and remakes itself, like the computer systems of some science-fiction space craft. This danger requires that we be constantly alert to the need to specify agency.

Power, authority and the firm in social systems of organisation of the construction process (SSOCPs)

Firms

Work in the tradition of Williamson (*Markets and Hierarchies*, 1975) draws a sharp distinction between authority-relations within a firm and contractual-relations between a buyer and seller. On the other hand, Alchian and Demsetz (1972) attempt to construct an explanation of the firm without recourse to a concept of authority.

Not only do we reject Alchian and Demsetz's position, but we go beyond Williamson's early position, to find elements of hierarchy in inter-organisational or inter-firm contractual relations. In particular we believe that such authority exists, within SSOCPs, in relations between clients and the professional service firms they contract with; and also, between main- and sub-contractors.

We also follow Marglin (1974, 1984) in emphasising that the existence and size of the firm cannot be explained solely, or even mainly, in terms of efficiency (minimum technically efficient scale of production), but rather in terms of power (control over the labour process; real rather than formal subsumption of labour to capital) on the one hand, and some of the 'laws of competitive accumulation of capital' (concentration and centralisation; excess of rate of profit over rate of interest) on the other.

One way of expressing the former of these two points is to say that the capitalist firm is the most efficient solution (to the problem of minimising the sum of production and transaction costs) only if we take as given the initial distribution of property rights. Another is to say that we only arrive at the capitalist firm as the necessary efficient solution if we rule out the possibility that workers can hire capital. It is perhaps also relevant however, to stress that real competition by no means ensures the survival only of the most profitable forms of organisation, and nor are efficiency gains almost always or even usually captured as profit.

At the same time, only a methodologically inveterately individualistic economics could posit the individual owner of capital as the basic unit of a modern capitalist economy rather than see the capitalist firm as its defining social institution.

Markets

Are there such things as markets in the construction process? If so, what are their key institutions? Hodgson (1988) proposes a useful distinction, for any institutional economics, between markets proper and simple, atomistic acts of exchange. Not every exchange involves a market, nor does the existence of an exchange suffice to create a market. He defines markets as 'sets of social institutions within which a large number of commodity exchanges of a specific type regularly take place'. Markets involve institutions: legal rules, customs, practices, embedded in specialist market 'agents' and 'regulators' which exist to service that market, and increase its efficiency relative to atomistic exchange.

These sources of superiority of the organised market over atomistic exchange relate to the three successive sources of transaction cost (see Chapter 5, p. 123) each relating to a stage in any transaction process (Dahlman, 1979; Coase, 1960), these being:

(1) search-and-information costs prior to identifying the parameters of an exchange;
(2) bargaining and decision costs prior to decision to commit an actual exchange act;
(3) contract monitoring/policing/enforcement costs.

Thus we have in an organised market: bodies supplying *ex ante* price, product and buyer/seller information; bodies to assist in bargaining and decision making; and bodies for transaction monitoring and enforcement.

These bodies can be competitive private firms, or collective market bodies subscribed to by all as a condition of market entry, such as the Stock Exchange or Lloyds. These institutions within markets perform certain functions. They establish and publish prices. This provides key contextual information for parties to each single transaction within a market, which transactors without a market lack. They inform buyers of the identity of potential sellers (and vice versa) and even indicate the perceived or real qualities of the products and services on offer. In some cases, they may define and enforce quality standards by which product quality can be measured and described. These agents within markets can act to centralise and 'net' the physical movement of goods from suppliers to users, and store stocks whilst exchanges occur. They may also assist in the transfer of property rights in goods, supplying the terms and implicit context of contracts, via such means as private market 'rules' and customs, bodies of relevant common law, legal enforcement procedures and standard forms of contract.

Sometimes, as with Lloyds, these market making bodies will serve just one specialised market. Other times, as with say the *Financial Times, Exchange and Mart*, market research firms, or commercial courts, they will provide similar services to many markets.

Markets in the above sense do not necessarily require the existence of multiple transactions in standard products with uniform prices, although market efficiency is increased when this occurs. Thus, although they lack the publication of a standard market price for their products and services, housing and construction markets have a rich set of their own market-making and market-serving bodies.

For a market to exist, there must be a specific type of exchange, so that each of the multiple transactions shares features which enable its transactors to benefit from information about other transactions of the same type. This information allows buyers to switch from one supplier to another within a market and links products and services in terms of their cross elasticity of demand. In construction separate markets may even be said to exit, wherever the contractual forms and terms are significantly different. Under the same standard form of contract, for example, information on prices of contracts with 'fluctuation of price' clauses is of little use to transactors of a 'firm price' contract. If the difference is significant enough, this must lead to the creation of two parallel markets.

Market serving institutions in social systems of organisation of the construction process

Price information

Using this institutional approach to markets, let us examine an old question; namely, how many housing markets are there?

Some market serving bodies in construction publish and sell information. For instance, Spons, Laxtons, Wessex all publish builders' price books. The BCIS and the larger firms of quantity surveyors also publish information on construction prices and costs. This information is produced in two forms. First are the prices of resources themselves, mainly labour and materials, that enter generally into determining construction costs. Second are the weighted average contract prices for a number of projects that are judged by the various market serving organisations to be relevant and comparable. Because the contract data is grouped into project types, its publishers decide on the conceptual framework of what constitutes comparability. They therefore produce an implied model of what determines variation in project prices. As a result they supply knowledge, not raw data.

Other specialist bodies, especially building societies and estate agents, supply house price information. The DETR also publish such information.

Information is often produced in the form of (time series) price indices for construction and housing. Again these market serving bodies provide knowledge of movements in the general level of such prices, imposing their own conceptual frameworks.

The Builders' Conference shares amongst its members data on the identity of all bidders and their prices for individual contracts. The construction trade press also publish information on contract prices and the identity of the contractors awarded these large contracts. The trade press also publish data on the identity of clients, their consultants and data on project characteristics for projects about to be tendered. Such data is also now available on-line (*Tenders on the Web*).

Quantity of demand and supply information

Construction market research firms sell information on trends and forecasts of demand and supply for narrow segments of the construction industry, e.g. the demand for bathroom suites. Other firms sell market research information on the demand for housing. Information on leading indicators, such as planning approvals and housing starts, is widely used to give short-run predictions of new supply. In this way professional transactors can have some information about supply volume several months ahead. However, this general availability of market forecast information does not extend to every building product. Forecasts of supply and demand for some products are made by the monopolistic producers privately and often in-house.

Predictions of supply and demand in construction and housing markets are almost always generated under conventional assumptions that present real price levels, or present rates of price increase, will continue into the forecast period. These assumptions are made in the absence of effective futures markets in housing and construction, preventing the use of a market validated price expectation. (Futures markets in commodities, such as tea and coffee, allow firms to sell/purchase in advance of production in anticipation of future price levels.)

Quality of product information and quality regulation

The market serving bodies which produce information on quality can be divided into producer firms themselves or their agents, such as advertising agents, and independent sources. Advice on quality is provided by bodies such as the Building Research Establishment, which produce guidelines to producers. They attempt to convert 'current safety and non-failure thinking' among scientists into practical guidance to builders and designers. Recently, this has expanded to include guidance on energy efficiency. In construction markets there is perhaps an understandable tendency towards regulatory enforcement of minima rather than reliance on popular publication of information on quality to inform consumer choice.

As far as regulation is concerned, building developers are by no means free to determine their own choices of quality, subject only to *caveat emptor*. Statutory building control inspection applies, together with private insurer organised inspection (e.g. NHBC). Specialised regulation and law exists to monitor and control impacts on owners of contiguous structures. These regulations involve issues such as party-walls, rights to light and restrictive covenants. In addition there are regulations setting minimum standards in respect of safety, habitability and energy efficiency.

Building production is also widely perceived to have impacts on the wider community beyond the producers themselves, the future occupiers of the buildings produced and the owners of contiguous properties. These wider issues are grouped under aesthetic, amenity and environmental impacts. Aesthetic considerations are mainly taken into account only in designated areas, while amenity impacts are primarily concerned with the derived demand on local spatial infrastructure. Environmental impacts have become increasingly controversial as pollution has climbed up the political agenda. In principle the local authority planning system is designed to take into account and protect the interests of third parties affected by construction projects.

All of these amount to quite a panoply of public quality control. In addition, actors in the process may choose to introduce private quality control agents, such as defects-insurance companies and surveyors. All these control agents are key market serving institutions.

An agenda for future research into social structures of housing provision

This chapter has raised a number of unanswered issues regarding social structures of built environment provision in general. For example, how has or how could development of new market-serving and market-making institutions: (a) increase the efficiency-advantage of organised markets relative to integration or atomistic transaction; (b) restructure the market in the sense of defining new sub-markets?

Conversely, how has the failure of certain kinds of market-serving institution to emerge reduced the efficiency advantage of markets relative to integration or atomistic transaction?

A number of hypotheses are also suggested pertaining to the social structures of housing provision in particular:

- that housing-asset (house ownership transaction) markets differ fundamentally from financial asset markets for the following reasons: lack of a 'model' on which any actor would be able to form 'rational expectations'; divergent expectations amongst transactors; lack of a 'futures' market; heterogeneity of the asset. We see the need to test alternative Austrian and post-Keynesian hypotheses about expectation-formation and subjective valuation, against the observed (using quantitative research methods) and reconstructed (using qualitative methods) behaviour of both developers and owner-occupiers.
- that changes in the spatial structure of the dominant SSHP can only occur if accompanied by changes in its social structure of relationships between actors.
- that consumers and other investors in the dominant SSHP have abandoned a set of 'certainties' about the appropriate long-run expectations and, therefore, behaviour in the private housing market that held across cyclical fluctuations and other shocks through the 1970s and 1980s.

Concluding remarks

We have seen in this chapter that the particular institutions of the social systems of organisation of the construction process are many and diverse.

The market institutions involve players and rules and arrangements for transactions. The players are the firms (principals and their agents), while the rules are reproduced in part through specialised market-making/market-serving institutions that do not participate as principal players.

Notes

1 National House Builders Council (1998), *New House-Building Statistics*, tables 13–15.
2 'If the price of any durable or storable good is to remain even briefly at rest, it needs to have attained a level which divides the potential holders of this

asset into Bulls and Bears of its price. For if all of them suddenly come to think alike, at that moment the price will change abruptly' (Shackle, 1972; p. 112); and, '[t]he market cannot solve the problem of expectation. The only price it can distill, for a storable .... good, is one which divides the potential holders of that good into two camps, those who think its price will rise and those who think its price will fall. For at the moment when all are agreed on the direction of the next movement, that movement, paradoxically, will prove to have already taken place' (ibid., pp. 83–4).

3

Size Distributions and Polarisation of Construction Firms and Markets

Introduction

The belief of many casual users of statistical data is that statistical distributions always show a central tendency, so that the average figure provides the best summary of the data. This is, however, not always the case in construction. The proposition of this chapter is that in fact, if anything, there are often spreads in the data which lead to bi-modal distributions. In other words, in some areas there are tendencies towards polarisation. These tendencies can be explained to some extent by the social, political, and cultural environment in which firms trade, the social structure of accumulation (see Chapter 1, social structure of the economy; social structure of production).

In this chapter we consider several examples of this duality in the construction industry: in the distribution of work carried out by different sized firms; in productivity data; and in the dual labour market. We also see the duality of petty and real capitalist sectors operating in construction.

The misleading average – the significance of different ways of looking at statistical data

Most economic data is collected in the form of many cases or observations of one variable. For example, the ONS and DETR collect data on the size of construction firms, as measured either by turnover or numbers employed. Each firm is one observation, and the variable being measured is size of firm.

Often published data are merely an average or mean figure. In some cases, because of the way the data is collected, it is, even before processing, in the form of an average – for instance, average productivity of a firm and the average number of workers employed by each firm. These averages may then themselves be averaged, to produce a figure for output per worker for the whole industry, or for all firms of a certain kind.

Industry averages are certainly important and useful, especially when we wish to compare two industries, or to analyse changes in a variable for one industry over a period of time. However, averages necessarily conceal all evidence of variation of single cases about the mean. If this variance, or dispersion, of cases about the mean is relatively small, there is no great harm done, but the greater the dispersion the more misleading can be the single mean figure.

One obvious example concerns average income or wealth per head in a country. The distributions of both income and wealth are highly dispersed and asymmetric with respect to the mean. A large number of people, mostly in households where no one has a job, have low incomes, perhaps below one-quarter or one-third of the national average. A larger number of people, perhaps the majority, have incomes somewhere near, say between half and twice, the national average. A substantial number have incomes several times the average; and a small number have incomes very many times the average. If income were represented by height, and the average person was around 5 feet 9 inches tall, the richest few would be over a mile high. Except for those who fall through its safety net, a welfare state will ensure that the poorest will at least have a height measured in feet rather than in inches. The average income per head measured by the arithmetic mean for the UK tells us very little about the likely income of any individual we might select at random. The tail to the frequency distribution, comprising people with very high incomes, pulls up the average. The lack of a corresponding tail at the other end of the distribution also helps push it up, so that the majority of people will have incomes below the average, and an income at the arithmetic mean might be quite unusual and in many ways unrepresentative. In this situation, we can have recourse to such single measures as the mode or the median. We can also describe the frequency distribution by quartiles or deciles, or indeed we can represent it as a graph.

In the non-economic world, many variables, such as height in the human population, are found to display what is called a normal distribution. This, when graphed, gives a bell-shaped curve, symmetrical about its mean, and with the mean coinciding with the mode and the median. More exactly, in a normal distribution two-thirds of all cases lie within plus or minus one standard deviation distance of the mean. The larger the value of the standard deviation relative to the mean, the more dispersed is the distribution.

Now, in the economic world normal distributions are rare. If they do occur, it will be purely by accident. Skewed and asymmetrical distributions, with their humps or peaks to one side of the mean, are perhaps more common, as we saw with income distribution.

The distribution of size of construction firm follows a highly skewed pattern. The smallest size range of firms, with 1 to 3 employees, occurs with the largest frequency. As the number of employees in each size range increases, the number of firms per size range declines. However, if we look at the shares in total turnover or employment taken by all the firms in each size range, we find something quite different. Table 3.1 shows a total of approximately 86,000 1-person firms; 48,000 firms with 2–3 persons employed; 16,000 with 4 to 7 employed; and, at the other end of the range, under 600 firms employing 115 or more; out of a total of just over 160,000 firms.

Now, in this table the size ranges are not all of the same width. If they were, one would probably find a steady tapering away of the number of firms as one moved to successively higher size ranges. This general effect is in any case clear from the table as it stands.

Suppose we arbitrarily decide to group all firms into just three size ranges, large, medium and small, and that these shall be defined as 1–7 persons employed, 8–114 persons and 115 persons and above. We then

Table 3.1   Work done by size of construction firm, 1997

Size range, by number employed	Number of firms	% of all firms	% share of value of work done	Cumulative % share of value of work done
1	86,269	53.8	10.1	10
2–3	47,644	29.7	9.5	20
4–7	15,737	9.8	7.9	28
8–13	3,787	2.4	4.9	32
14–24	3,101	1.9	7.1	40
25–34	1,176	0.7	4.8	44
35–59	1,156	0.7	7.5	52
60–79	396	0.2	4.3	56
80–114	296	0.2	4.8	61
115+	586	0.4	39.1	100
All firms	160,148	100.0	100.0	

Source: Tables 3.1 and 3.3 in DETR (1998), *Housing and Construction Statistics 1987–97*.

see that 28 per cent of all work was done by the small firms, 33 per cent by the medium firms and 39 per cent by the large. If we then bring time into the picture, and compare this with the distribution of work done by these three size ranges of firm in earlier years, we are in a position to test a theoretical hypothesis – the hypothesis of the disappearing middle, as it might be known. Figure 3.1 illustrates that in the period from 1979 to 1997, there was a slight trend for the share of the middle band to diminish, though the tendency varied from year to year, with 1996 even showing a possible temporary reversal.

Polarised distributions, or ones which tend to become more polarised over time, are of special interest. They suggest that in economics, the mean and median of the distribution of a variable may not always be the most representative or characteristic statistic. It is sometimes more insightful to think of a variable not as a continuum from one end to another along an axis with an average somewhere in the middle but in terms of opposite poles, each with their own 'magnetic force'. It has been observed, for example, in many less developed capitalist economies that the economic structure tends to polarise into, at one pole a small number of large monopoly or near-monopoly producers,

Figure 3.1   Percentage of work by size of firm 1979–97: the disappearing middle?

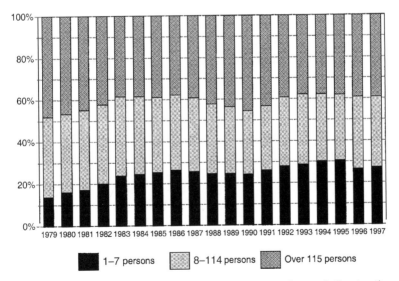

Source: Table 3.3 in DOE/DETR (various dates), *Housing and Construction Statistics*.

often either foreign or state-owned and usually protected by or benefiting from state economic policy, and at the other a vast number of petty commodity producers, too small to be classed as capitalist enterprises at all, and operating in the informal economy outside of legal regulation or recognition, with rather little in between in the way of medium-sized enterprises.

Over the very long run (for example, over the twentieth century as a whole) we think, the size structure of the British construction industry is tending to polarise, into an ever larger number of small firms, with a growing share of total work done, and a diminishing number of ever-larger firms at the other end of the scale, with a stable share of total work done, resulting in a squeeze on the total output-share of middle-sized firms, and above all a reduction in the total number and average size of such firms.

There are some good *a priori* reasons for expecting this to happen. On the one hand, small repair and maintenance projects account for a growing proportion of output and very small firms are at an advantage in this market over medium sized firms. The growth of subcontracting also favours the emergence of more small firms. On the other hand, the industry's biggest firms seem able to outgrow the industry as a whole, partly because of faster internal growth from competitive advantages over medium sized competitors and partly because of external growth through take-overs and mergers. Medium-sized firms, meanwhile, are not as flexibly adapted to very small projects and sub-contracts as small firms. Nor can medium-sized firms take advantage of the economies of scale and range of resources of large firms.

Economic polarisation: barriers to equalisation

More generally, economic polarisation may arise due to virtuous and vicious circles. Both economic success and failure may be self-reinforcing and therefore cumulative. An example can be found in regional development. High growth regions may indeed experience labour shortages, rising wages and congestion costs, whilst low-growth regions experience unemployment and falling wages. But, instead of these factors working to restore the relative balance of economic advantage or attractiveness to capital investment of the two types of region, investment tends to concentrate only in the growth regions. There the rate of productivity increase tends to be highest and so does the rate of growth of local markets and hence demand, especially for services, for which local demand is often the most important source of demand.

Lower wages cannot compete with higher productivity and higher demand to attract investment capital away from the booming region to the depressed one. Thus one region continues growing whilst another continues to decline, without obvious limit. The economic rationale of regional policy in the European Union is to attract investment into peripheral low growth areas through state aid.

This principle of cumulative causation is the direct opposite of the belief that market forces will tend to bring about an equalisation of returns and a balanced equilibrium. In the conventional, neoclassical view, parts of the world economy may differ in their resource endowments, but, unless market forces are prevented from operating, both economic success and failure will be self-limiting. That is, economic growth will itself tend to drive up costs and wages in a sector by creating shortages. Economic decline will tend to drive down costs and wages. If labour is immobile between sectors, capital will then flow from more successful sectors to those where it can obtain higher returns or profits as a result of their lower wages.

If factors of production are immobile between sectors, then if factors in a sector are in excess supply and partially unemployed or underutilised, because there is insufficient demand for the sector's products relative to its capacity, these factors' prices must drop until this creates an extra demand for them, by making that sector's products cheaper to produce. Provided factor prices are fully flexible, the theory suggests that factor unemployment will not be permanent. Likewise, in sectors with excess demand, factor prices will be driven up, by scarcity, thus pushing up product prices until they rise sufficiently to choke off the excess demand for factors, which is of course a derived demand, and which in turn depends on the demand for those products. In this case the same factors will command very different prices in different sectors, but each sector will find its own full employment equilibrium.

If on the other hand factors are mobile, then they will flow from sectors where their supply exceeds the demand, i.e. where their prices are falling, and flow into sectors where demand exceeds supply and factor prices are rising. These flows will occur until an equilibrium is reached in which the same factor commands the same price in whatever sector and factor prices are stable, when flows will cease. In this kind of economy some sectors can grow continuously whilst others shrink, so long as the shrinking sectors contain, and therefore can release, enough of the factors to sustain growth in the other sectors. Essentially, over the long run factor quantities in each sector adjust, and factor prices are relatively stable and uniform.

In reality, a whole range of technical, political and legal factors to do with the system or regime of regulation of the economy determine whether consumer goods, capital goods, money capital, technology and labour are able to flow freely between economic sectors in which their present prices or returns are different. A different result can be expected, for instance, if goods are allowed to flow freely but capital and labour cannot, than would be the case if capital and labour could also move. Maddison (1991) has developed a system of periodising the economic history of the advanced capitalist economies (ACEs) using just these criteria.

In certain periods or phases consumer goods, capital goods, money capital, technology and labour are highly mobile. This constitutes, according to Maddison, an economic regime most supportive or generative of economic growth in *successful* regions or countries. One such phase, he suggests, was the 'golden age of capitalism', between the completion of the recovery from the crisis after the end of the Second World War, roughly 1950, and the crisis of 1973–4. In this phase the economic system or regime was one which placed minimum barriers to the international flow of goods and reduced barriers to the international flow of labour (migration) and money capital. At the same time, barriers to the transfer of technology were reduced, at least between the USA, the technological leader, and other ACEs. In other phases, the system or regime is less propitious for growth – here Maddison cites particularly the period from 1914 to 1950 with its restrictions on trade, migration and capital movements. To a large extent such a belief in the economic benefits of the free flow of goods, services and labour is behind the introduction of the Single European Market and its extension to more countries in the future. In essence the Single Market creates a new economic regime with new opportunities for growth. It is important to note that Maddison's model, and data, concerns only the ACEs. It does not deal with countries which have not developed, and does not imply that these countries too benefit in the same way from an international regime of low barriers.

Maddison's concept is essentially one of barriers, or their absence, to inter-sectoral flows of capital, labour, technology and commodities in a world economy. Maddison himself is most concerned with differences in growth rates between national economies. He therefore divides the world economy into national sectors, distinguished by significant gaps in productivity, derived from differences in economic structure, such as the proportion of the labour force in high productivity activities, technology in the highest productivity activities, capital stock per

Table 3.2 System characteristics of different phases in the economic growth history of the advanced capitalist economies

	Government policy on unemployment/price stability	Nature of international payments system	Labour market behaviour	Degree of freedom for international trade	Degree of freedom for international factor movements
1870–1913 Liberal phase	no concern with unemployment	gold standard with rigid exchange rates	weak unions; some downward flexibility of wages	very free; no quantity or exchange restrictions; tariffs the only barrier	more or less complete freedom
1913–1950 Beggar-your-neighbour phase	concern with price/exchange rate stability leads to conscious acceptance of large scale unemployment	gold standard restored at obsolete parities; 1931 system collapse followed by movable peg	governments enforce downward wage flexibility	quantity and exchange restrictions widespread; tariffs raised substantially	severe controls on both capital and labour
1950–73 Golden age phase	priority given to full employment	fixed (but not rigid) exchange rates with large international credit facilities available to governments	strong unions; no downward wage flexibility	very strong move towards freer trade and customs unions	gradual but substantial freeing of both labour and capital movements
1973– Cautious objectives phase	priority given to price stability	system collapse followed by floating rates and growing areas of stability (EMS)	weakened unions	free trade maintained	freedom for capital movements augmented; labour movement restricted

Source: Maddison (1991), pp. 120–21.

worker, and proportion of work force with education and training to the levels found in the highest productivity activities in the leading country. If barriers to inter-sectoral flows (in this case, international flows) are high, then inter-sectoral economic performance will diverge, with the result of a widening gap between the poles. However, all of Maddison's arguments are generalisable to any multi-sectoral model of a whole economy, so long as there are potential barriers to flows across sectoral boundaries. Borders between nations are merely the most obvious example of a more general phenomenon. Sectors in this sense only exist if there are social institutions which, as it were, police and reproduce their boundaries. In other words the different economic sectors must be social realities and not mere statistical abstractions.

To give a simple, if extreme, historical example of barriers *within* one national economy – an economy in which a *free* urban industrial sector coexists with agricultural serfdom has such high barriers to flows between these sectors that a widening polarity between them will result. Rural serfs, being compulsorily tied to the land, cannot move to add to the labour force of the urban economy. The result is likely to be an ever widening gap between living standards and levels of productivity in the rural and urban sectors. Emancipation of the serfs would then have an effect analogous to the removal of barriers on international flows in Maddison's model.

Polarisation and duality in the construction labour market

Using the labour market as an example, let us now try to bring together the two ideas explored above and others from Chapter 1. Firstly, the economy consists of socially defined sectors with barriers to flows between them; and secondly, the economy is underpinned by a social structure of accumulation.

In certain social structures of accumulation (SSA) there are magnetic forces at work that tend to reproduce each pole of the structure, whilst at the same time tending to prevent any convergence. For example, the SSA of the US economy between 1950 and the 1970s has been defined by some economists in part by its containing a *dual* manual labour market – a high-wage, white, male, unionised class of workers at the one pole, and, at the other, its opposite, a low-wage workforce of black and female (and non-unionised) workers – linked to a separation between high and rapid productivity growth industries, mostly using Fordist mass production methods to make consumer durables, on the one hand, and low and slow productivity growth industries on the other hand, mostly producing services. The socio-economic structure of the period contained barriers, concerning race and gender, which made labour mobility between these sectors much harder and less than mobility between jobs within either sector. It is this lack of intersectoral mobility that in fact defines sectors, and makes them an economically meaningful concept.

Sometimes, the relationship between economic sectors poles apart can be symbiotic. The high wage and high productivity sector of the economy, for example, may be able to concentrate most of the available supply of investment capital precisely because the low wage and low productivity sector has low capital investment requirements, yet can absorb all the increase in the national total labour force. This prevents mass unemployment of a kind which could destabilise the SSA. The high productivity sector cannot expand employment as rapidly, since it cannot generate many more jobs precisely because its productivity is growing as fast as its output. However, when an SSA reaches crisis point, the relationship between sectors or poles becomes contradictory. Developments or events in one sector start to impede accumulation of capital in the other sector, or start to threaten the social stability of the whole structure. An example would be when, in the 1970s, workers in the low paid sector in the UK began to demand pay increases to catch-up with standards of living in the high pay sector, while workers in the latter responded with demands to restore customary pay differentials.

It can therefore be seen that for a dual labour market to exist, there must be at least two conditions. Firstly, there must be a polar distinction between two types of jobs, each type possessing opposite attributes, and secondly, there must be barriers preventing all or some workers moving from one kind of job to the other, more attractive kind. Two important models of the segmentation of jobs have been proposed.

In one model, the division is between manual or clerical jobs and non-manual professional (including technical) employment. Professional jobs compared to manual jobs are held generally to possess attributes of flexibility in work rules, and relative autonomy from direct supervision or managerial task design – that is, they show a high amount of independence with which workers execute their tasks in the labour process. These are known in the literature as independent jobs, and contrasted with subordinate ones.

In the other model, the division is between core or primary and secondary jobs. Here the distinctions concern levels of job security, pay stability, wage form and career progression. Primary jobs offer higher levels of all of these attributes than secondary jobs. The primary jobs tend to be in large establishments, and in large, oligopolistic firms or government agencies, and are most likely to be unionised. They can, in principle, be any jobs in such employers, including manual or clerical ones. Secondary jobs tend to be in small firms in the competitive sector; or, if the employer is a large firm, the jobs are part-time or on a contract of fixed duration. The jobs of self-employed workers and labour-only subcontractors fall in the secondary sector.

The sectoral division of workers within the contemporary UK is most clearly between genders and ethnic groups, as seen for example in average pay differentials between men and women, or in differences in unemployment rates between ethnic groups. These gender and ethnic divisions of workers are reflected in the types of people employed in different industries and at different grades. There is also an argument that formal education serves as a device to mark and perpetuate a social class division of the work force. Education and professional qualifications may erect *artificially* high barriers between an educated social class and a relatively uneducated social class. There are also important regional divisions and barriers to mobility.

In gender terms, construction is very clearly part of the male sector of the labour force. Taken as a whole, the construction industries have one of the lowest female participation rates to be found in the economy. In ethnic and national citizenship terms, however, we find that the UK and EU construction industries draw quite heavily upon ethnic minority, immigrant or non-citizen work forces. Construction activity is, of course, spread across all regions and draws upon all regional work forces. During the 1980s construction cycle of boom and slump the proportion of total new orders in the South East and Greater London grew and then contracted. This still leaves open the question of the degree of inter-regional labour mobility and wage differentials to be found within construction.

There has also recently been a clear tendency for direct recruitment into the administrative, professional, and technical parts of the construction labour force quite separate from the operative side, and for less individual movement across this boundary. This may have coincided with a widening relative pay differential between these two sectors, though the evidence is made obscure by difficulty in estimating average LOSC operative pay (Ive and Gruneberg, 2000).

The sectoral division of capital in the modern UK is perhaps strongest between the *small business* or *petty capitalist* and *corporate* sectors, with significant barriers to the flow of money capital or technology between them.

The duality seen in the construction labour force reflects duality found in other industries and sectors throughout the economy. The divisions caused by the social structure of accumulation tend to run through the middle of the construction industry. It is therefore not helpful to characterise all of construction as ever belonging to one pole or another of the overall dual structure of the economy. Rather, the economic and social divisions in the construction industry represent a microcosm of society.

In the rest of this chapter we concentrate upon two dimensions of duality in the modern UK economy and construction industry:

- that between the high and low productivity sectors of *production*
- that between the corporate and petty scales and forms of *business organisation.*

Productivity differentials in construction

An economic duality occurs in the construction industry related to productivity and wages. In one sector relatively high wages are accompanied by high productivity, while in the other, firms compete by cutting wages, finding the cheapest sources of labour supply, while apparently paying less attention to trying to raise productivity. This division in the construction sector can be seen using statistical evidence from official censuses of construction firms.

Although it has been argued that the available productivity and wage statistics for the UK construction industry are not particularly reliable, well devised conceptually or consistently defined, they nevertheless provide a limited insight into differences between the high wage/high productivity and the low wage/low productivity parts of the industry.

Two sources of statistical data can be found, one in DETR's *Housing and Construction Statistics* (HCS) and the other in ONS's *Census of Production* (COP). Chapters 1 and 2 of *Housing and Construction Statistics* contain data on aggregate construction output and employment of labour, whilst Chapter 3 contains information on productivity by trade and size of firm. The *Census of Production* gives figures for productivity and wages by size of firm and NACE classification. *Housing and Construction Statistics* covers Great Britain and the *COP* the UK, though this cannot be a significant source of difference between them.

Table 3.3   Productivity differentials – method 1: private contractors' productivity: HCS chapter 3

	1991 Value of work done per head per annum: (£ current prices)	1991 As % of all-firms average productivity	1997 Value of work done per head per annum: (£ current prices)	1997 As % of all-firms average productivity
Firms employing 1–24	31,250	73	43,290	78
Firms employing 25–114	48,400	114	65,290	117
Firms employing 115+	57,600	135	70,160	126
Building and civil engineering main contractors	61,320	144	76,480	138
General builders	40,640	95	57,070	103
Civil engineering contractors	60,840	143	61,860	111
Specialist trade contractors	36,540	86	49,380	89
All firms	42,640	100	55,610	100

Source: Calculated from Private Contractors' Census, 3rd qtr, 1991 and 1997, Tables 3.3 and 3.4 in DOE/DETR (1992; 1998), *Housing and Construction Statistics.*

Notes:

(1) annual productivity estimated by multiplying 3rd quarter productivity figures by four;

(2) employment = total employment = operatives + APTC + working proprietors; thus, in principle, excluding self-employed LOSC workers; total employment (according to this data source), 1991 = 866,600; 1997 = 778,500

(3) total (annual) value of work done by contractors (according to this data source) in 1997 was approx. £43,000 million; and total employment by contractors was 778,000; these totals must be compared with those computed from other sources before making comparisons between these and other productivity estimates (see below);

(4) there are discontinuities and differences in coverage between 1991 and 1997.

The most striking feature of Table 3.3 is the large differential it shows in productivities in firms of different sizes. Also, whilst we know that size of firm correlates with type of firm (specialist trade contractors on average in 1997 employed just over 4 persons each whereas building and civil engineering contractors on average employed nearly 20 persons each), only a small part of the size-related productivity differential can be explained as the result of differences 'innate' to types of firm regardless of size, together with association of size of firm with type of firm.

Direct comparison of the above data with those in the *Census of Production* (COP) is made difficult by the fact that the latter uses somewhat different size categories.

The Census of Production distinguishes the following categories of firms: larger undertakings, defined as those employing 20+ people, broken down into eight size ranges and five classes according to their chief type of activity; and all smaller firms, employing 0–19 people, grouped together in one class. One of the classes of larger firms is itself divided into state-owned and privately owned firms. Thus we are presented with data for seven types of firm in all, and for nine size ranges.

The Census of Production does not obtain returns from smaller (0–19) firms. Instead, their ratios are estimated, using the results for small firms' output and employment in DETR/DOE's Private Contractors' Census (PCC).

The latest COP from which data had been published by the time of writing was that for 1995 (ONS 1998). Our data is based on 1991 and 1995 data. Table 3.4 shows the derivation of the gross output per head, based on the value of gross output divided by the number employed; whilst Tables 3.5 and 3.6 show net output or value added per head.

Several points of detail in Table 3.4 require clarification. Employment refers to employees (APTC and operatives) and working proprietors. Thus, in principle (as with method 1, above) self-employed LOSC workers are excluded. In 1991 total employment (according to this data source) was 1,035,239 and in 1995 it was 967,700 persons. These employment totals are significantly larger than those derived from Method 1 (see notes to Table 3.3, above). For example, in 1991 the difference is 169,000 workers. Some 121,000 of that 1991 difference is accounted for by inclusion of public sector DLOs in Method 2 and their exclusion in Method 1.

Total gross output in Table 3.4 refers to value of work (including subcontract work) done and work in progress and includes the sale of goods and receipts for services rendered to other organisations (includ-

Table 3.4 **Productivity differentials – method 2: all undertakings (private and public sectors): gross output per head: COP**

1991	No. of firms (number)	Employment (£m)	Gross output (£m)	Gross output per head (£)
Undertakings employing 1–19 (estimated)	200,941	418,260	19,149	45,780
Undertakings employing 20+ of which:-	5,296	617,980	47,399	76,820
NACE 500	705	137,970	4,397	31,870
NACE 501	2,141	222,210	23,571	106,080
NACE 502	697	113,070	11,659	103,120
NACE 503	1,040	101,620	5,618	55,280
NACE 504	713	42,120	2,154	51,140
1995 Enterprises employing 1–19 (estimated)	181,406	473,600	33,397	70,500
Enterprises employing 20+	5,854	494,100	45,800	92,690

Source: Calculated by authors from data in CSO (1993), *Report on the Census of Production 1991, PA 500: Construction industry*, and ONS (1998), *Business Monitor PA 1002, Production and Construction Inquiries, Summary Volume 1995*.
Notes:
NACE 500 = general construction
NACE 501 = construction and repair of buildings
NACE 502 = civil engineering construction
NACE 503 = installation of fixtures and fittings
NACE 504 = building completion.

ing hiring-out of plant, etc.). It is in principle identical to the standard definition of gross output in economics (see Ive and Gruneberg, 2000; ch. 1) and in the Blue Book. COP reported gross output in 1991 was £66,548 million, compared to the value of work done per year in Table 3.3 above, from PCC, of £36,950 million. The latter is, evidently, somewhere between a measure of net output or value added and a measure of gross output.

The reporting unit for the COP in 1991 was the *undertaking* but the definition of an undertaking, according to the COP 'need not be a single geographical location'. By 1995 the reporting unit definition had changed to the *enterprise*, which the COP defines as 'the smallest set of legal units with a degree of autonomy within an enterprise group'.

Finally, NACE 500 consists chiefly of local authority DLOs. These have an output, value added and productivity profile that is very different from that of private contractors. For present purposes, of comparing private firms of different types and sizes, this row is probably best ignored.

Table 3.5 shows the share of gross output paid for materials and sub-contracted work. This leaves the value added by contractors. What Table 3.5 brings out most clearly is the fact that there are two types of private construction firms. Type 1 are main contractors who receive orders direct from clients and subcontract-out much (well over 50 per cent) of their gross output (value of orders; turnover; sales) to other construction firms. Type 1 firms have high levels of gross output per employee (see Table 3.4), but low ratios (around 20 per cent only) of value added to gross output. Type 2 firms are sub-contractors who receive most of their orders from main contractors and in turn subcontract-out some of this to yet other firms, but a relatively lower proportion (around 25 per cent). Type 2 firms have low levels of gross output per employee.

Type 1 firms are mainly found in NACE 501 and 502, while NACE 503 and 504 contain mainly Type 2 firms. However, comparison with numbers of 'main' and 'specialist' firms in Table 3.3 makes it clear that many specialist trade firms (mostly Type 2 sub-contractors) are grouped together with Type 1 firms in NACE 501 and 502 – thus making these NACE categories less useful for economic analysis than one would hope them to be.

Since the COP does not directly gather data on small firms, but relies for such firms instead on data supplied by DOE/DETR, one can only recommend extreme caution when comparing COP 'estimates' for small firms with COP census returns for larger firms. Broadly speaking, this weakness in the COP data makes us prefer to use the PCC data (method 1) as a source for *comparing* productivity in smaller and larger firms – even though COP data may well have value as an estimate of the *absolute* level of value added and value added per head in larger firms.

The productivity gap and size of firm: conclusion

The Census of Production and PCC figures for productivity and wages per head are strongly suggestive of an industry divided into a low wage

Table 3.5   Shares in gross output of value added, materials and subcontracted work in 1991

	Gross output (£m)	Subcontracts as % of gross output	Materials purchased as % of gross output	Value added as % of gross output	Value added by undertakings (£m)	Value added as a % of gross output
Undertakings employing 1–19	19,157	32.0	36.4	31.6	6,049	31.6
Undertakings employing 20+	47,383	46.7	27.4	25.9	12,257	25.9
of which:-						
NACE 500	4,403	18.1	29.8	52.0	2,289	52.0
NACE 501	23,554	57.0	23.1	19.9	4,686	19.9
NACE 502	11,678	50.3	27.0	22.7	2,646	22.7
NACE 503	5,614	26.9	39.4	33.8	1,897	33.8
NACE 504	2,154	25.4	40.3	34.3	739	34.3
All undertakings	66,520	42.5	30.0	27.5	18,306	27.5

Source: Calculated by authors from data in CSO (1993), Report on the Census of Production 1991, PA 500: Construction industry.

Table 3.6 Productivity differentials – method 3: all undertakings (private and public sectors): net output (value added) per head: COP

	Value added (£m)	Value added as % of gross output	Employ-ment	Value added per head	Wage costs (£m)	Wage costs per head (£)	Wage costs as % of value added
1991							
Undertakings employing 1–19 (estimated)	6,049	32	418,300	14,460	5,550	13,270	92
Undertakings employing 20+	12,257	26	617,000	19,870	9,872	16,100	81
All undertakings	18,306	28	1,035,200	17,680	15,440	14,920	84
1995							
Enterprises employing 1–19 (estimated)	11,225	34	473,600	23,700	6,065	12,810	54
Enterprises employing 20+	11,724	26	494,100	23,730	8,605	17,420	73
All enterprises	22,949	29	967,700	23,710	14,670	15,160	64

Source: Calculated by authors from data in CSO (1993), *Report on the Census of Production 1991, PA 500: Construction industry,* and ONS (1998), *Business Monitor PA 1002, Production and Construction Inquiries, Summary Volume 1995.*

Notes:

1. Value added in 1991 and 1995, at current prices

2. In principle, in both years, data for 'wage costs' are for 'wages and salaries' and 'employer's NI contributions', and should exclude all payments to working proprietors.

3. However whereas the data for 'employment' include working proprietors (WPs) in both years, by 1995 payments to WPs are excluded from 'wage costs', whereas in 1991 such payments are included: 'Estimates of wages and salaries and NI contributions for undertakings with an employment of less than 20 are based on total employment, including working proprietors', note b to Table 3, CSO (1993).
It is hardly possible that, in 1991, wage costs of employees *other than working proprietors* accounted for 92% of the value added of small firms (the ones with working proprietors), since this would leave a negligible amount of value added to be distributed to a large number of WPs. Even allowing for negative profit incomes in that year, it seems certain that significant WP income must indeed have been included as wage costs in that year. On the other hand, it is equally implausible that, in 1995, 'wage costs' *including payments to working proprietors* accounted for only 54% of the value added of small firms.

4. Value added is normally estimated in two ways, by deducting purchases from gross output *and* by summing reported labour and capital incomes. This permits cross-checking. Where incomes are estimated, as they are in the COP for small firms, the resulting figure for value added must be regarded as less reliable.

with low productivity sector and a high wage with high productivity sector, along lines of size of firm and its correlate, type of work.

In periods of high demand at least, LOSC workers, for whom we do not have direct data, probably extend the high wage, high productivity pole still further. If, in fact, most LOSC workers work for the larger firms, then this would confirm and widen the productivity gap between large and small firms. The larger firms show much higher productivity than the smaller firms.

The scope for higher real wages and living standards in the low wage segment of the construction industry from a redistribution of existing value added between profits and wages appears to be limited. Gross profits of smaller private firms comprised only 8 per cent of their value added in 1991, admittedly a recession year (though to this we must add quite a large allowance for profit income disguised within the total income of working proprietors). This figure is highly suggestive of a sector of tiny firms operating on the narrowest of margins between net output and wage costs, and without substantial financial reserves. For firms in this precarious situation, falls in output require immediate downward movement in wage costs. They therefore resort to devices to control and reduce wages, in response to drops in output, that can take effect in the very short term. This means downwardly flexible wages per worker, flexible levels of employment per firm, and concentration on methods for squeezing extra productivity such as raising intensity by exhortation and threat. The working proprietors' and other heads of tiny firms' own income depends directly upon their success in this. They may reduce the porosity of the working day (Ive and Gruneberg, 2000, Chapter 3) by changing terms and conditions of employment and shifting the cost of unproductive time onto the worker. Alternatively they may lengthen the time worked per worker. We would expect these methods of raising productivity to have more limited effect in boom periods, so that productivity increase in this sector, such as it might be, would occur mainly in recession periods.

However, the biggest source of possible productivity increase, and potentially the most mutually beneficial to workers and firms, comes from improved efficiency. This however requires resources of time and space, in the sense of a gap between immediate output and unavoidable outgoings. Small firms in construction work on credit from builders merchants and sub-contractors for work in progress. To execute a strategy for improving efficiency, it is necessary to devote resources to a stream of initiatives which are, from the perspective of the firm, experimental. It also requires a certain scale of operation. Because many ideas to increase efficiency involve investment, they involve a fixed, indivisible expenditure (if only of time and managerial effort) to be recouped by marginal benefits to each unit of a sufficient volume of output, and the elapse of time between outlay and reward. Investment involves uncertainty and risk, with at best only a probability of success. In our view, most construction firms are simply too small and lacking in capital and other resources to meet these requirements.

Sources of higher productivity in the large firms sector

How then do the larger firms obtain their higher productivity and faster rate of productivity increase? One explanation that can be largely discounted concerns the hypothesis that it is because they own or invest in larger amounts of plant and equipment or other fixed capital per worker (see Table 3.7).

Ownership and use of fixed capital are not the same thing of course, and it remains possible that larger firms hire or lease a larger proportion of the fixed capital they use, compared to smaller firms. Unfortunately we have no figures for expenditure on plant and equipment hire and leasing disaggregated by firm size, to test this.

Other possibilities are that larger firms achieve lower porosity of the working day, higher work intensity, or greater non-capital-embodied efficiency. We will explore one of these possibilities in depth, below. We have seen that the larger firms pay higher wages per worker. The argument is that it is in the larger firms that we find an implied bargain that workers will work with above average intensity and in return will receive above average wages.

Productivity of firms and type of work

What small private firms and DLOs have in common, as well as relatively low productivity, is that they both mainly engage in works to existing structures (WES), and especially repair and maintenance.

Table 3.7   Capital investment per person per year at current prices

Size of firm, by employment	1993	1994	1995
1–19	1,040	1,630	1,540
20–499	956	1,282	760
500+	446	886	730

Source: Annual Censuses of Production and Construction, PA1002, Summary Volumes 1993, 1994, 1995.

Figure 3.2   Composition of work by size of private contractors, 3rd quarter 1997

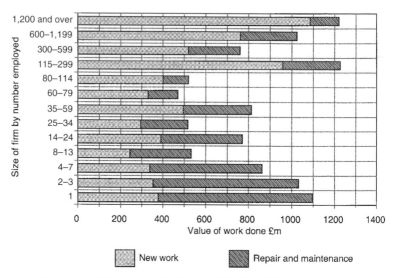

Source: *Housing and Construction Statistics, 1987–1997*, Table 3.8.

Figure 3.2 shows how the ratio of R & M to new work varies inversely with size of firm. In 1997, R & M was much the main source of work (64 per cent) for all firms employing up to 7 persons. It was roughly half of output (51 per cent) for firms employing 8–24, only 22 per cent for firms employing 115–299, and as little as 11 per cent of output for firms employing 1,200 and over.

We do not know how many worker/years of labour expended within a firm were on R & M work, and how many on new work, and thus cannot directly measure this productivity differential. However, the association of a high proportion of R & M in the output of a category of firms with relatively low productivity is so clear and consistent that we can be in little doubt that a multiple regression analysis would be able to 'explain' a substantial proportion of productivity difference between firms by use of a 'R & M share in output' variable.

It is equally easy to construct an *a priori* explanation of why we would expect to find lower productivity in R & M work: smaller projects, less repetition, more difficult site conditions (e.g. less scope for access for plant and equipment; need to 'work-around', and limit disruption to, occupants), need to accommodate to dimensions and placing of existing building elements, etc.

Market power and productivity

The larger-firm sector appears to show much higher levels of productivity – as much as 50 per cent higher, on some measures in Tables 3.3–3.6. However, before we rush to attribute all of this difference to different project characteristics on the one hand and to superior efficiency, economies of scale or a different version of wage-effort bargain on the other, we must consider one other possibility. The difference in measured performance between large and small firms in construction may be partly due to the exercise of market power.

Insofar as the larger firms stand to the smaller ones in a relation of main to subcontractor, the terms of trade between the two will influence how much, if any, of the value of the output of the latter can be appropriated by the former. The lower the prices at which subcontractors sell their output, the higher will the net output of main contractors appear to be.

Indeed, if we introduce a distinction between directly productive workers and others (responsible for directing and supervising the former, for setting them their tasks and for selling and collecting payment for their output, but not directly producing tangible, sellable output themselves), we might go further and suggest that the employment of the larger firms is disproportionately of non-directly productive workers. As a proxy and test for this, in 1997 (DETR, 1998; *HCS*, Tables 3.11 and 3.12) the ratio of APTCs to operatives in firms employing 115+ was 0.65 to 1, whereas in the industry as a whole it was 0.47 to 1, excluding LOSC. Indeed, firms employing 115+ together employ nearly half (48 per cent) of all APTC staff in the industry, but only just over one-third (35 per cent) of the operatives.

Firms in 'main contractor' categories in 1997 employed APTC to operatives in the ratio 0.58 to 1, whereas for firms in 'specialist trade' (subcontractor) categories the ratio was 0.38 to 1. Specialist trade firms employed 53 per cent of the industry's operative workforce, but only 42 per cent of its APTC staff.

We can regard the net output of the industry as being produced only by productive workers so long as we recognise that part of the work of APTC staff is concerned with the organisation of the work of others, and thus impacts upon their productivity. *Productivity per productive worker* is then, of course, much higher than productivity per worker of all kinds. However, while most directly productive workers are employed by the smaller, subcontractor firms, much of the total value added produced in the industry is attributed to, and appropriated as gross incomes by, the larger, main contractor firms.

The extent to which APTC staff employed by main contractors can and do organise the work of operatives employed by subcontractors then becomes the relevant issue. If we assume that the extent to which this can be done effectively is limited, then we must consider that a significant proportion of all APTC staff employed by main contractors must be engaged in other kinds of activity – either in work arising from the need to control transaction costs (rather than production costs) or in efforts to control the terms-of-trade between main contractors on the one hand and both clients and subcontractors on the other.

Accurate estimates of the number of directly productive workers in each part of the industry would require that, in addition to directly employed operatives, we had accurate figures for LOSC workers, and fully understood the balance of activities of working proprietors.

In 1997, working proprietors, who own their business, manage it and work in the firm themselves while often also employing others, numbered 162,000 (DETR, 1998; *HCS*, Table 3.13). If, say, half of their time is directly productive work, they input the equivalent of 81,000 operatives.

In 1997, private sector operatives numbered around 420,000 (418, 000 if taken directly from the PCC; 384,000 plus some part of the 112,000 estimated employees not working for firms on the register of firms, if derived from Table 2.1 in HCS); and LOSC (estimated as all self-employed in construction, 593,000, minus working proprietors, 162,000) numbered 431,000. The total private sector productive labour was therefore approximately equivalent to 1,013,000 people out of a total of all private sector construction workers of 1,294,000 (adding-in 200,000 APTC staff plus the other half of the time of working proprietors).

Capitalist and petty capitalist organisation sectors: polarity of profitability, accumulation and competition

A further duality exists in construction between real capitalist firms and petty capitalist producers as illustrated in Figure 3.3. Real capitalist firms we propose to define as limited companies, owning at least £60,000–£80,000 of their own capital employed, and employing at least 14 to 20 persons or more. Fourteen employees coincides with the Private Contractors' Census borderline for size of firm while 20 is consistent with Census of Production statistics. In 1991 there were 7,100

Figure 3.3 The petty and real capitalist sectors

firms in the 14 and over category in the Private Contractors' Census, and 5,300 in the 20 and over category in the Census of Production.

The petty capitalist sector we propose to define as comprising *near-firms*, run by self-employed working proprietors, employing 3 or less, and *micro-firms*, employing more than 3 but less than 14 or 20.

While both real and petty capitalist firms require financial capital in order to function it is the established real capitalist firms which have a ready access to funds, finance and credit and can borrow at a lower rate of interest than small firms with working proprietors. Moreover, because of their size and dominant position in construction markets, it is the capitalist firms which are better able to exploit the benefits of large plant and equipment embodying better technology than that available to smaller firms with working proprietors.

In this way micro-firms and working proprietors survive on low profits while failing to accumulate capital. This petty capitalist sector persists for several reasons. It is in the interests of capitalist firms to employ small competing low price subcontractors. There are also markets, such as small domestic repair and maintenance work, in which larger firms find it unprofitable to compete. Moreover, petty capital cannot directly flow into the capitalist sector to obtain attractive corporate rates of return. Whenever a small firm does find a way of making high profits, it is usually bought out and absorbed into the capitalist sector. Thus there is a persistent difference in both the cost of raising capital and in the average rate of return on capital employed in the two sectors.

There are virtually no barriers to entry in the petty capitalist sector. In 1991 according to the Private Contractors' census shown in Table 3.8,

Table 3.8 Private contractors' census breakdown of employment type by size of firm, 1991

Size of firm	No. of firms	Working proprietors	Operatives	APTCs	All employment	WPs as % of total
1	103,200	94,500	nil	nil	94,500	100
2–3	70,500	83,500	46,200	17,800	147,500	57
4–13	26,600	30,200	81,200	30,300	141,700	21
14–24	3,400	2,900	42,600	15,500	61,000	5
25+	3,700	1,600	267,500	152,800	421,900	0.4
All firms	207,400	212,700	437,500	216,400	866,600	25

Note: WP = working proprietor.
Source: Calculated from Private Contractors' Census, 3rd qtr, 1991; Table 3.3 in DOE (1992), Housing and Construction Statistics.

over 170,000 firms employed just 1 to 3 persons, including their working proprietors. Some 70,000 of these *near-firms* were general builders or contractors or engaged in main trades, such as bricklaying. The other 100,000 were in specialist trades. Each year there is a rapid turnover of near-firms on the DETR's register. Over the business cycle a large number become bankrupt or cease trading but an equally large number start up. From 1982 to 1991 some 15,000 construction partnerships and sole traders became insolvent and were made bankrupt by their creditors, as did 19,000 companies, the overwhelming majority of these being very small firms.

Many self employed working proprietors in construction, often employing others as well as working themselves, earn no more, over the cycle, than the average LOSC worker, hired by a larger contractor, selling only their labour and working for wages. We can learn something about these near-firms by looking carefully at the *Private Contractors' Census* tables in *Housing and Construction Statistics*. Here total employment is broken down into employment of operatives, APTCs and working proprietors. We see that 74 per cent of total employment in near-firms was of working proprietors (Table 3.8).

Since there were some 70,500 firms in the 2–3 employment category, it follows that the average such firm had 1.2 working proprietors, 0.65 operatives and 0.25 APTCs. That is, most had just one working proprietor and employed one operative or APTC, though a significant minority may have had two working proprietors, but employed no operative or one operative or one APTC. The micro firms employing 4–13 and numbering 26,600, contained on average just 1.1 working proprietors, 3 operative employees and 1.1 APTC employees.

The difference between the micro-firms and the near-firms is that the labour force of near-firms, even allowing for some employment of LOSCs by them, consists mostly of the working proprietors themselves, whereas in the micro-firms the ratio is such that it is likely that their working proprietors concentrate mostly on entrepreneurial, managerial and supervisory work.

Building repair and maintenance work is also home to a substantial number of franchise businesses (Felstead, 1993) Those operating under franchises, too, may perhaps best be called near-firms. They certainly lack the autonomy of operation of a fully fledged capitalist firm.

Most near-firms never make it even to micro-firm status. This may be partly a matter of deliberate intention. Often their owners aspirations are only to be autonomously self-employed, and they have no desire, nor do they believe they have the opportunity, to become owners of small capitalist firms. Even more to the point may be the low productivity levels of their firms – too low, in fact, to permit them to employ others profitably at prevailing average levels of employment cost.

Additionally there are high barriers preventing micro-firms from entering the sector of real capitalist firms. Lack of own financial capital (start-up capital plus accumulated retained profit) and access to bank and trade credit are two such barriers. Another is created by many clients' and management contractors' procedures for approving firms for selective tendering, which reject firms without past experience of contracts of similar size. This goes alongside lack of informal contacts with clients. Another is lack of experience in certain key business areas, such as estimating and tendering, tax, and cash flow planning and management, expertise in which is perhaps more essential for the larger limited companies in the real capitalist sector than the near- and micro-firm.

Construction is a highly competitive business for the petty capitalist sector near- and micro-firms, selling an undifferentiated product with competition largely based on price. Nevertheless most micro-firms manage to make satisfactory rates of profit on turnover and on capital employed, even during recessions, often by depressing wages. They are able to make sufficient absolute gross surpluses over employment costs to yield sufficient capitalist incomes for their owners, comprising a mixture of salaries, directors' fees, and benefits in kind, such as private use of company owned assets, as well as declared profits. These capitalist incomes are 'sufficient' to convince the owners to keep the business going, but insufficient to allow accumulation of capital by retaining profits in the business.

It is striking that, even in depressed demand conditions, fully fledged capitalist construction firms rarely, if ever, try to enter the market preserves of the petty capitalist firms, suggesting that they are well aware of the less favourable conditions for profit making that apply there.

Concluding remarks

We have seen in this chapter several examples of variables such as the size of construction firms that do not follow a normal distribution. They may be heavily skewed, as in the distribution of income, or bi-modal as are various aspects of the labour market.

We have explored ideas of polarisation between firms (and workers) of different types, and that socially produced barries may exist which prevent movement between these poles, despite the exitence of more attractive conditions at one pole than at the other. The nature of any such barries will vary between periods with different social structures of accumulation. Each social structure of the economy, and each social system of organisation of the construction process, throws up its own characteristic polarities and barriers.

Firms located at one pole then operate under very different conditions and in a different context to firms at the other pole. Accordingly each type of firm faces completely different pressures, and must select its business options from different possibility–sets.

Part II
Construction Markets

4
Towards Usable Concepts of Construction Market Structure

Introduction

Having looked in detail at housebuilding markets in Chapter 2, we now expand the discussion to construction markets in general. We begin by elaborating on some aspects of the conventional industrial economics model of the market, such as market structure, concentration, and barriers to entry. From a seller's point of view we discuss price elasticity of demand applied in a construction market context, and we conclude the chapter by discussing market segmentation and marker specialisation by construction firms.

In this way it is possible to adjust the conventional model of the market to at least partly account for some aspects of market behaviour in real construction markets.

Construction markets

A conventional definition of a market is given by Devine *et al.* (1985), who state that a market exists wherever buyers and sellers of a *homogeneous product* are in sufficiently close contact with each other that a *single price* prevails. However, it is questionable whether this strict definition of a market can be applied in the context of buildings and built structures, because in reality, we do not find here a homogeneous product or service with a single price.

The idea of a product market was originally developed to cover analysis of standard, homogeneous commodities, such as agricultural or mineral goods. In this sense, corn or coal are products, and indeed we may find corn and coal markets, organised as Exchanges with licenced dealers and market makers. The buyers and sellers of corn at

Casterbridge Corn Exchange (in Thomas Hardy's *The Mayor of Casterbridge*) are clearly 'in sufficiently close contact ... that a single price prevails'. Homogeneity of the good and closeness of contact of the actors are both necessary assumptions for this concept, but so is the idea of a highly specific set of market institutions to handle information.

The concept of the market is derived either from the theory of a commodity or the theory of the firm. The theory of perfect competition places emphasis on the commodity whereas monopolistic competition theory emphasises the role of the firm. The concept of *product-market* links the microeconomic concept of the market to the mesoeconomic concept of industry. The concept of industry refers to a sector or division of the whole system of production comprising an economy. If we assume that each firm produces just one product, then we can define an industry as the set of firms using the same specific types of resources of labour, equipment and technology. The supply side of a market is then a set of firms producing a product. The output of the firms and the quantity of the product available will be identical. That is, on these assumptions, an industry equals a market, and thus the two terms become substitutable.

In reality, of course, many firms produce in more than one industry, whilst markets consist of multiple segments. Thus even within an industry firms sell in many separate markets or market segments.

For the first half-century of the history of neo-classical economics the term *market* actually referred to an homogeneous commodity, in the above sense. The *market economy* was seen as a set of commodity markets. Equilibrium in each market was consistent with equilibrium of the whole economy. This neo-classical approach to the interaction of markets within an economy is known as the model of general equilibrium. However, Sraffa (1926) pointed out a major contradiction in this theory.

Sraffa's point was, in part, that under the perfect competition model of price taking firms facing horizontal demand curves, the profit-maximising level of output per firm was indeterminate, unless firms faced long run average cost curves that eventually rose with volume of output. But perfect competition theory was unable to explain why one firm's long-run cost curve should rise in this way. If optimum firm size is indeterminate, there is nothing to stop one firm monopolising a market, or at least obtaining a significant market share. However, if the condition for perfect competition that many firms each produce only an insignificant part of total output of a commodity is abandoned, then this renders invalid the whole perfect competition model of price determined independently of the actions of each firm. Each firm is no longer a price taker with the total output volume unaffected by a firm's output decision, nor is each firm able to choose the quantity it sells by finding the quantity at which its marginal cost equals a given market price.

Following Sraffa's criticism of perfect competition, Robinson (1933) and Chamberlin (1933) reconstructed the theory of the single market in equilibrium by abandoning the notion of general equilibrium in the whole system. Their emphasis shifted from the determination of the equilibrium level of a commodity's output and price, towards the decisions that would put the firm into an equilibrium position. A firm would be in equilibrium when in an optimum situation, where profit is maximised at a certain level of output. Once in equilibrium a firm would have no reason to move or change. This will be the output at which the marginal cost equals the marginal revenue. Demand and supply conditions from now on would work indirectly to determine output and price for a commodity or product by influencing the cost and revenue schedules, and hence the decisions, of firms. However, Robinson and Chamberlin performed this theoretical rescue operation in ways which opened the door to the concept of product differentiation.

Once we allow each firm some scope for active decision making, then we must allow that one of the decisions it can make in order to maximise its profits is deliberately to differentiate its product from those of rival firms. However, product differentiation means there are no longer clear boundaries between commodity A and B, each with its own market equilibrium price. Instead there is a range of differentiated products of single firms, with each firm setting its own price and deciding its own output volume. In such a world, the meaning of *markets* undergoes a profound loss of clarity.

Differentiation in construction markets presents a rather different problem to the more usual text book example of product differentiation by producers. Under traditional contracting systems, or even sometimes under atomistic non-market transactions, it is clients, with their designers acting as agent, and not the producer firms, who specify and control the form and content of the built product. This has shifted the focus of attention from the product to the type of service undertaken by contracting firms. In turn this has led some construction economists to describe the *product* of the construction firm not as the building but as a construction *service*. In this way it is possible to use the now conventional concept of differentiation (e.g. Hillebrandt and Cannon, 1990), to point out differences between firms competing for the same project in terms of differences in the service provided, although the final building might be identical in terms of its appearance, regardless of the contracting firm employed. Nevertheless, the rules of selective competitive tendering on price contain an assumption that all tenderers are equivalent or undifferentiated in terms of the quality of the service they are offering. Differentiation under this market arrangement is confined to a simple one dimensional sorting of firms into sheep and goats by the customer – into approved and non-approved, or tender listed and non-listed, firms.

Market structure

A substantial body of economic analysis uses the concept of a market mainly in order to derive a theory of market structure and of how that structure affects the behaviour of firms within a market.

By *market structure* is normally meant *inter alia* the degree of concentration or distribution of market shares, which is the proportion of all transactions in a product involving each buyer and seller. Market structure is also concerned with the extent of product differentiation. By producing a distinct product or service a firm is in a position to protect its own sales from changes in the prices of other firms, to some degree. Hence the price elasticity of demand experienced by a firm in response to a change in its own price or in the price of its closest market rivals is a feature of market structure. Indeed the price elasticity of demand for the product as a whole is a major determinant of market structure since the behaviour of producers will be dependent on the reaction of customers. Finally, the height of barriers to new firms entering the market is a feature of market structure since producers are aware of the potential threat posed by an increased number of competitors.

If we follow the neo-classical approach and retain, for the moment, the assumptions that firms are single product profit maximisers, then it becomes fairly straightforward to derive inferences about behaviour from factors affecting the shape of the cost and revenue curves that a firm faces. For firms to maximise their profits the neo-classical rule is to select the combination of price and output at which the rising marginal cost curve cuts the falling marginal revenue curve. Market structure then matters to the firm in so far as it affects the shapes of those curves.

For instance, in perfect competition firms are assumed to be price takers. As they have no control over the price they charge, it is up to the firms themselves to ensure that they select a volume of output such that their marginal cost of the last unit is equal to the market price, in order to maximize profits. Moreover, producers in perfect competition make no attempt to distinguish their products from those of their competitors.

In contrast to this theoretical perfectly competitive market is oligopoly, in which a few firms, which dominate the market, are obliged to enter into a variety of non-price competition, such as advertising, promotions and corporate imaging. Just as in perfect competition, but for different reasons, in oligopolistic competition price is rarely used as a weapon to win market share. Price competition between oligopolists would be destructive. If one firm undercut the price of its rivals, they would retaliate with price reductions of their own. Such price wars would only lead to lower prices without necessarily increasing a firm's share of the market and the threat of this outcome is sufficient to deter firms from lowering their prices to increase their market share.

Instead firms in oligopolistic competition attempt to differentiate their products and services. In imperfect competition, firms differentiate their products and price them above or below their perceived competitors, taking quality and marketing differences into account. Here the attempt is to acquire a niche in the market, a gap not covered by existing products or services provided by competing firms. The purpose of firms acting in these ways is to create a situation in which direct comparison between products and services is difficult. As a result, it is possible for firms to increase the gross margins of their sales by taking advantage of the perceived added value of a given branded product. In order to achieve this they must introduce imperfections into the market and product differentiation is one of the methods used.

An alternative strategy which oligopolistic firms may adopt is collusion. By agreeing, tacitly or explicitly, overtly or covertly, oligopolists can determine price and output of each producer and the allocation of work and market shares. This enables oligopolists to charge a higher price than would otherwise have been the case, if competition between them had been allowed to erode profit margins. In fact, the higher prices may act to cushion the higher costs of firms which would otherwise be forced to become more efficient. There is evidence that oligopolists are therefore sometimes operating at higher prices but at profit margins which are similar to those in more competitive markets.

In monopolistic markets, the dominant firm controls the market to such an extent that it is possible for it to determine either the price of the goods and services or the volume of goods and services sold. It cannot determine both price and quantity. Nevertheless, the conclusion that can be drawn from neo-classical theory is that it is in the interests of monopolists to restrict output, raise price and earn supernormal profits despite producing at a level of output below the minimum point on the long-run average cost curve. In this way the firm maximises its profits without producing at the most efficient level of output.

At the same time, monopolists and oligopolists are in a position to take advantage of segmentation by charging different customers different prices, even for the same product or service, according to the customers' ability and willingness to pay. This form of marketing is known as price discrimination and is widely seen where firms can separate customers and prevent onward selling from one market segment to another. Thus, telephone charges at different times of day, travel by train by different types of individual (students, elderly, business travellers), electricity prices to industrial and domestic users, are all forms of price discrimination. Price discrimination enables monopolies to increase sales revenues by offering lower prices to some and higher prices to others. This particular market structure allows this kind of pricing behaviour provided there are no close substitutes, as far as the customers offered the higher prices are concerned.

Against this, we now compare and contrast a set of alternative approaches, which depart more or less fundamentally from the neo-classical framework. We will examine, in turn, market concentration, barriers to entry and price elasticity of demand.

Concentration: seller concentration

For any product or service, given a state of technology, there is a level of output per firm below which unit costs begin to rise. This level of output is known as the minimum efficient scale of production (MES). The larger the MES, the bigger the average size of firm in a market, since firms with outputs below the MES will not be able to compete in the long run, as their unit costs will be higher than the industry norm.

In the neo-classical view, seller concentration is presumed to reflect the MES. The larger the MES relative to the total size of the product market, the higher the degree of seller concentration that will maximise industry efficiency. An industry is said to be efficient when the total resources used to produce a given output is at a minimum. It is assumed that intense competition between firms will force an industry structure towards this efficient structure. At any other structure, the opportunity exists for firms to make increased profits by changing their scale of production. No other structure meets the requirement of *equilibrium*.

Thus, for industry equilibrium:

$$n = Q/MES \qquad (4.1)$$

where
n = number of firms of identical, cost-minimising size
and $\quad Q$ = total size of the product market

Hence,

$$MES = Q/n \qquad (4.2)$$

Seller concentration is measured as the proportion of total sales per firm. Assuming all firms are of identical, cost-minimising size, then:

$$C = (Q/n)/Q = 1/n \qquad (4.3)$$

where
C = seller concentration

Seller concentration is used as a proxy measure of the shape of long run cost curves and reflects economies of scale in an industry or market, since

$$MES = Q \times C \qquad (4.4)$$

and the value of output at the *MES* is the minimum cost per unit of output.

However, there is another, completely different way of looking at seller concentration and its effect on the behaviour of firms, which is derived ultimately from Kalecki (1971). The higher the market shares of the few largest firms, the higher the *degree of monopoly* and the higher the mark-up they will be able to add to average costs. Where there are only a few dominant firms, tacit or explicit agreement to avoid price wars is more likely. As a result prices will be maintained.

The higher customary mark-up will be protected by the more powerful threat of retaliation that dominant firms can make, either against new entrants or price-cutters. Certainly, much management literature takes it as axiomatic that dominant market share is a desirable strategic objective for a firm, because it will permit the firm to obtain sustained higher profit margins (Buzzell and Gale, 1987). There is also much empirical research suggesting that, above a certain threshold, market share is positively associated with profit.

In practice we find that firms of widely differing sizes co-exist in most markets. If the largest firms reap cost advantages from economies of scale, yet set prices equal to those of the smaller but higher cost producers, then their profit margins will be higher. Alternatively, the dominant firms with large market share may use their position to increase the differentiation of their product lines. They may, for instance, spend relatively more as a percentage of turnover on developing new products or on advertising. This would also have the effect of giving them higher gross profit margins and probably higher net margins, while also increasing their volume of sales still further.

For construction, the key question may be not the overall industry seller concentration ratio but whether firms, of whatever size, have small numbers of known close rivals. If construction markets are heavily segmented by factors such as project size, project technology and location, then, for instance, medium sized firms specialising in specific markets need not necessarily face lower specific market shares or more close rivals than the largest firms. It would be with these rivals rather than with the overall industry leaders that their behaviour would be inter-dependent.

Concentration: buyer concentration

Sellers behave differently if they face a fragmented population of many purchasers than if they are selling to a small set of buyers. In new speculative housing markets, for instance, there is a multiplicity of buyers each with a negligible market share. Construction capital goods markets, on the other hand, are characterised by a high buyer concentration, dominated by small numbers of large corporations or public authorities. In broad terms, the higher the degree of buyer concentration, the less will be the scope for high margins and high profitability among sellers. We would therefore expect lower profit margins among contractors producing large capital projects than amongst sellers of new housing.

Apart from the number and size distribution of buyers, several institutional factors concerning buyers come into play, each potentially having an effect on buyer and seller behaviour, and ultimately on seller profits. For instance, buyers may combine to exchange information about their projects, about sellers and about themselves. They may act jointly to develop standard procurement practices and forms of contract. They may be bound in their behaviour by sets of rules, such as codes of procedure for selective tendering and, in the public sector, *standing orders* designed to provide transparency, equity and accountability, and to check corruption.

Barriers to entry

Protected by barriers to entry, firms in a market will tend to set their margins and prices jointly at higher levels than they would if they feared the arrival of new entrants. Tacit market sharing behaviour and co-operation between producers may also occur. High barriers to entry bring stability to sets of rival firms. They increase each firm's degree of confidence that it can predict the behaviour of competitors and of the industry at large.

All existing firms will have a clear common interest in keeping prices and margins high, and in *managing change* in an orderly way. For example, they will wish to control the rate of obsolescence and replacement of expensive fixed capital equipment in accordance with *predicted* or planned levels. Competition between established rivals is conducted with the aim of keeping up with, or only modestly out-performing, the industry average. The ambitious new entrant, on the other hand, must build up a market share from scratch, and will seek to introduce and benefit from increased uncertainty and unplanned change. 'Pocket electronic calculators are more likely to be introduced by a new entrant to the mathematical-instrument business than by established major producers of slide-rules', nor was it the leading Swiss precision watch makers who introduced digital watches (see also Schumpeter, 1934). Barriers to market entry therefore play a key role in standard industrial economics, both in the shorter run where the emphasis is on prices and margins, and in the longer run where the emphasis is on innovation.

a   Economies of scale

Barriers to entry exist in terms of costs and in terms of access to potential customers. On the cost side, barriers are associated with the concept of the *minimum efficient scale* of production, below which it would be uneconomic to set up in competition with existing firms. The MES defines the minimum investment required of a new entrant, if it is to be cost competitive.

On the demand side, barriers are associated with the concept of customer loyalty to existing suppliers or to their brands. To some extent these barriers to sales are therefore a consequence of the degree of advertising intensity, which is the proportion of total market sales spent on advertising. They will also reflect expenditure on quality management systems, designed to keep customers satisfied with the quality of a firm's products.

b   Supply chains

Most construction markets are characterised by low advertising intensity and by low expenditure on quality management as well as low production-side economies of scale. This implies that barriers to entry to construction markets are very low and, in turn, that rates of profit for existing suppliers are also low. However, before drawing such conclusions it is important to note that other barriers to entry can be established by vertically integrating the activities of existing firms. Alternatively long term contractual arrangements are often established with other firms in the production supply chain. Such arrangements can form effective barriers to new firms. For instance, existing petroleum refiners or car manufacturers may have petrol or vehicle retailers tied to them by systems of dealerships. A new entrant to these industries would therefore have to invest capital to establish its own network of retailers. This would increase the cost of the total investment needed to enter the market beyond that required simply for efficient production. An example of this type of barrier to entry in construction is road building contractors who may own all existing quarries for road making materials. A new entrant contractor would then have to open its own quarries.

c   Incumbents' cost advantages

The conventional idea of how barriers to entry work rests on a notion of firms' investing in creating production capacity, but this notion is not directly applicable to construction. In the conventional model of manufacturing or distribution new entrants plan to seize a share of a constant total market from existing suppliers. However, if the new entrants have to set up their own production and distribution capacity, they are likely to increase the problem of excess capacity for all firms, including themselves. Their entry adds to total capacity whilst leaving total demand unchanged.

Established firms are protected from competition by new entrants if the latter will be unable, on entry, to match the former's level of costs. This is likely to be the case where experience and learning-curve effects are strong (Jackson, 1982). However, as Salter (1969) has pointed out, if there is continuous technological improvement in production methods, existing producers may be left with obsolete, higher cost plant. The firm that enters by setting up new capacity then has the advantage of being able to install the least cost, most up-to-date technology, leaving aside questions of scale. In construction, apart from building materials production, it is doubtful whether most acts of entry by setting up new supplier firms actually add to total industry capacity. Instead, new suppliers plan to *poach* resources piecemeal from those firms already in the market.

d   Private Information

Private information can be the most powerful of entry barriers. Simple text book analysis tends to assume *perfect knowledge.* All actual or potential producers are assumed to have access to the same technical and market knowledge. In reality, however, existing producers may possess either private or proprietary knowledge. With private knowledge the firm is in a position to take advantage of information not known by other firms. With proprietary knowledge the information itself is publicly available, but its use for commercial purposes is protected by ownership rights, copyright and patents. It is, by and large, the private kind of information impactedness that is characteristic of construction. Private information about markets and demand can be just as important as technological information impactedness. Established firms may have knowledge about customers, subcontractors and competitors that new entrants will not have. This information is not shared and lack of it forms a barrier to entry.

Any technological or market information not generally available outside a firm will actually be known by certain individuals currently employed by that firm. The firm itself is not a sentient entity, nor does it own the individuals who do have the knowledge. It is therefore open to a new entrant to try to acquire this private knowledge by poaching those employees. However, existing firms may try to protect themselves from this threat by either restricting internal access to key information to a few personnel, and then rewarding these staff for loyalty and tying them to the firm by special contracts, or by fragmenting knowledge so that no one individual owns enough of the whole *jigsaw* for their knowledge to be of independent commercial value. Moreover, even if the knowledge itself is not the *property* of a firm, any document or disc onto which it is written will belong to the company. Thus poached members of staff can, in law, take with them only what they can memorise.

This kind of problem is particularly severe for firms following strategies of 'relational contracting'. Over time, the firm builds up commercially valuable close relationships with particular customers. However, in practice it is normally particular staff who have that relationship and the private knowledge of, and access to, the customer that it implies. If these members of staff leave they may well be able to take their customers with them.

e   Client imposed barriers to entry to contract construction markets

In construction there are negligible barriers to entry at the bottom end of the industry. Those barriers that do exist higher up are largely client imposed. It is these barriers that are key to the theory of growth of the firm in the construction industry, and to the structure of particular markets. However, because bottom-end barriers to entry are so low they do not limit the total number of firms in the construction industry and its overall structure.

Growth of a contractor can be modelled as a series of steps against size and complexity of projects undertaken. Within each step the firm grows quantitatively by taking on more projects of the kind it is already *approved* to carry out and in which it has experience. As a result, it encounters no client imposed barrier to access to this kind of project. However, since each such project market is limited in size, and faces increased numbers of suppliers resulting from market entry from below, eventually the firm wishing to continue to grow must move up across the next step. At this point it faces a *Catch-22* situation, that clients may sensibly only shortlist tenderers who can demonstrate past experience on similar projects. However, by definition, this is just what the firm lacks. Such barriers can be overcome, largely by taking advantage of overlap between different client views as to relevant project size bands. Nevertheless, one major limit on the growth rate of construction firms is clients who insist on employing only experienced firms, especially when a high degree of complexity is perceived to apply to a project.

On the other hand, projects involving innovation, originality or a lack of close precedents may permit firms to persuade clients to consider them on the basis of the strength of their ideas or methods, since no particular past experience will be deemed highly relevant, in the sense of permitting the experienced firm simply to repeat what it has done in the past. Indeed, in one sense, because of the uniqueness of each construction project, it is usually the case that some construction design firms market themselves on the basis of solving new problems with new clients, while others market experience with similar projects. (Winch and Schneider, 1993).

f   Contestable markets and contract construction

A certain set of trading conditions may create a *contestable market*, in which existing firms modify their behaviour in order to deter increased competition from potential new entrants. Incumbent firms act *as if* potential suppliers would be attracted by, for instance, above average profit margins in a particular market. Thus, firms see a market with potential entrants as a contestable market. In Baumol's (1982) model of contestable markets, entrants can immediately replicate the costs of any incumbent firm. New entrants establish themselves before incumbent firms can make any price response. Markets are contestable if there are no *sunk costs* and entry is perfectly reversible, so that a firm could leave the market and recoup all its costs of entry. *Potential suppliers* are all those firms who could, if they wished, enter a market and become active suppliers. They include all those who have, or could acquire, any specialised knowledge of techniques or of demand, plus the capital to enter at the required scale.

Measuring construction market shares and barriers to entry

Given a tightly defined type of project and small numbers of customers, it is feasible to measure the number of firms *in* a market by analysis of client lists of approved or short-listed suppliers. One such study is reported in Ive (1983) concerning the market for new public housing projects in London in the 1970s. This study used the concept of *potential* suppliers and distinguished between potential suppliers in the short and the long run. Approved lists of contractors were held by the local authorities acting as customers. Selective competitive tendering meant only those contractors on approved lists could tender. From time to time new firms had the opportunity to apply to join a borough's standing list of approved contractors for particular types and value ranges of work. In the long run, therefore it was open to non-listed firms to try to enter the list. In the short run, however, membership of such lists was fixed. Thus, short run potential supply could only come from the set of approved firms.

'For the London boroughs as a whole, there were well over 1,500 firms on their approved lists of main contractors ... Twenty-five per cent of all main contractors and general builders with addresses in London and included in the Private Contractors' Census ... were on these approved lists (1000 out of 4000); as were five per cent of all such firms registered in the rest of the South East ... More significant than the total approved lists, however, are the (top) contract value panels ... which (covered) virtually all new housing contracts. Only ... some 450 firms were on these relevant contract value panels.' (Ive, 1983).

These 450 then constituted the short-run potential supply for the market as a whole, subject to the caveat that many were not on the approved list of all boroughs, and were therefore only potential suppliers to part of the market.

By no means all potential suppliers were active suppliers, with a positive share of the market. In the four survey years, Ive found 212 active suppliers, with an average of just under three contracts per firm. However, just three of these firms accounted for ten per cent of all contracts let. Thirty three firms accounted for 52 per cent of the market by value. The proportion of sales of the top five firms in given years is shown by the five-firm concentration ratios in Table 4.1.

Since the study found that it was largely different firms that were dominant in the inner and outer London markets, concentration ratios for each of these are also shown. It may be helpful to know that the London Boroughs' public housing market in the four years covered by the Ive (1983) study comprised 60,000 dwellings in contracts let, or an average of 15,000 dwellings per year.

Thus in this particular construction market, clearly defined according to type of client and project, there was a relatively high 5-firm seller concentration. However, it is plausible to argue that the conduct of the firms with leading market shares might have been constrained or

Table 4.1   Largest 5-firm concentration ratios

Year	All London	Inner London	Outer London
	(%)	(%)	(%)
1967	45.1	49.0	51.6
1970	32.1	48.3	39.3
1973	30.3	39.7	40.1
1976	25.1	36.8	36.6

influenced by their knowledge that a substantial potential supplier population existed even in the short term, given the number of firms which were inactive but eligible. That most potential suppliers did not enter can either be ascribed to their strategic assessment of the long run prospects of the market or to their assessment, whether accurate or not, of profit rates being obtained therein.

Now, the problem with finding out about profit rates in a specific market, in a world where each firm operates in many markets, is that the only publicly available information on profitability relates to industries or to firms but not to markets. Industry data is available in the Census of Production, from stockbrokers' industry reports, and sources such as FAME (a CD-ROM and on-line database of information from company accounts), while information on individual firms can be found in their company accounts. How then, if firms in a particular market were enjoying super-normal profits in that market, would potential suppliers become aware of this?

A method is set out in Ive (1983). Assuming we know the identity of all actual suppliers and the values of contracts won in a particular market, it is possible to produce a proxy estimate of market profitability. This can be obtained by weighting each firm's profitability by its share in the market. For example, the unweighted average rate of pre-tax profit on capital employed for all firms active in the London public housing market in 1975–6 was 33 per cent. The weighted mean rate was 38.3 per cent, significantly higher than the unweighted average profitability of 21 per cent, given by Inter Company Comparisons (1977) for all UK medium and large contractors in that year.

Now, in 1975–6 firms with total turnovers of over £2 million accounted for 76.5 per cent of all contracts won in the London Boroughs' market, and therefore had a combined weight of 0.765 in the calculation of the weighted mean. The next question must be: how important was work in this market for these firms? The greater the proportion of its total turnover, the bigger would be the impact of this market on each firm's profitability. In fact, for contractors with turnovers of £2 million or more, the ratio of their share of the London public sector housing market to their total turnover was just over 20 per cent on average.

Let us imagine a representative firm with a turnover of over £2 million. Using the figures given above, as an average firm, 20 per cent of its turnover would have come from contracts in the London public housing market. Let us also assume that the firm made the average proxy market profit on capital employed on its work in this market, of

38 per cent. Finally, let us assume that it made the industry average profitability on the rest of its work of 21 per cent. Thus, 20 per cent of its work produced a rate of return of 38 per cent, and 80 per cent of its work produced a rate of return of 21 per cent. Its overall profitability would then be:

$$(0.2 \times 0.38) + (0.8 \times 0.21) = (0.076 + 0.168) = 0.244 \text{ or } 24.4 \text{ per cent} \qquad (4.5)$$

This formula makes the assumption that the firm allocates its capital employed between markets in proportion to its derivation of turnover. From the fact that the firm had 20 per cent of its turnover in this market, it is assumed 20 per cent of its capital was employed to support its work in this market. In fact, a firm's above average profitability in a given market may result either from of a higher than average margin of profit to turnover in that market or a lower than average ratio of capital employed to turnover, or a combination of both.

Comparing the figures on profitability given above, the unweighted average profitability for firms active in the London public housing market at 33 per cent was 12 percentage points greater than the average profitability of all firms in the industry of 21 per cent. Some 3.3 percentage points of that difference of 12 points could be explained by the super-normal profits available at the time in that particular market.

The method discussed here could in principle be used in order to estimate the average profitability of all contracts in a particular construction market, with a view to determining whether a firm should enter that market. However, a substantial market research or information cost would be incurred. Moreover, most of the relevant data is held by clients, who might deny data access to a potential supplier.

One might expect groups of clients to combine to undertake, or commission from an independent source, analyses of supplier profitability in their market. If the results were then published and profitability were found to be high, new entrants would be attracted. However, this has not been widely done, even by public sector clients for whom there is no commercial rivalry problem in pooling such information. This suggests that clients are not particularly keen to inform potential suppliers or market entrants of market conditions.

Alternatively, a contractor wishing to enter a market could analyse the published prices of winning tenders in a market against its own estimating department's hypothetical cost estimates for those projects. However, if the firm lacked the tender documents, these estimates could only be highly approximate. Moreover, at most, such analysis would produce an estimate of what the firm's *ex ante* profit mark-up on direct cost might be if it tendered competitively in the market, whereas the relevant knowledge on which to base the decision to enter would concern *ex post* profits. Analysis of published tender prices reveals nothing about the average gap between *ex post* out-turn and *ex ante* tender prices in that market.

Price elasticity of demand

Elasticity of demand is the responsiveness of the quantity demanded to changes in some other variable, including the price of the product itself, the prices of other products or the incomes of purchasers. In fact, the responsiveness of demand to any variable which could conceivably have an impact on sales volume can in principle be measured.

Price elasticity of demand is given as:

$$PED = \Delta Q / \Delta P \qquad (4.6)$$

where

PED = price elasticity of demand

ΔQ = change in quantity demanded

and ΔP = change in price ('own price' for a firm or 'product price' for a commodity)

The concept of cross elasticity of demand measures the responsiveness of demand to a change in the price of an alternative or a complementary good. For instance, cross elasticity of demand would show the relationship between the demand for one firm's output and the price charged by its competitors. Cross elasticity of demand is given as:

$$XED = \Delta Q_a / \Delta P_b \qquad (4.7)$$

where

XED = cross elasticity of demand

ΔQ_a = change in the quantity of a

and ΔP_b = change in the price of b

The concept of 'product price' elasticity of demand assumes that products are homogeneous. We immediately have the problem of whether it is possible to apply the standard concepts of an *homogeneous product* and hence *product market* in a construction context. In construction, *product markets* can be seen as sets of projects, clients and producers.

Suppose we use the empirical sub-divisions of construction orders or output available from *Housing and Construction Statistics*. This would define, for example, 'new public sector housing' as a product market. The idea is that all projects of this product type are closer substitutes for one another than would be projects of another type. Because these projects are spatially fixed it is recognised of course that they cannot be regarded *ex post* (i.e. after their construction) as substitutes in the eyes of clients. However, *if* the firms and resources supplying the market are mobile geographically, then the tender prices those firms offer *ex ante* should tend to converage (since each firm is offering to supply units of a similar, homogeneous product).

The more each client treats and values the product as if it were homogeneous, the more each client must regard offers to supply other similar clients with the similar product as close substitutes to the offers received from its own supplies. In such a market we would then expect to find a single prevailing normal price, around which there would be only very limited variance from project to project.

Now, if homogeneous product markets are going to have any actuality in construction, the public housing of the 1960s and 1970s would clearly be one of the best places to look. Not only were client objectives set mainly in volume terms, as the number of units built, but there were such product market institutions as the *housing cost yardstick,* which imposed standard construction price norms per person housed, in terms of *bed spaces.* Moreover, the customers of this market by and large refrained from trying to meet their requirements by purchasing from the existing stock, and thus we have a construction product market separated for once from the second-hand market in existing built stock. This market was supplied by smallish numbers of 'system builders' – i.e. firms offering proprietary prefabricated (mainly high-rise) concrete panel housing design and construction 'systems'.

What then do we find? In spite of a large degree of administered standardisation throughout the country by central government at the Department of the Environment, each local authority, acting as client, maintained its own list of approved contractors, so that few firms were on the lists of all clients. Even fewer firms, however, were active nationally. The supply side of this product market comprised sets of regional suppliers, and the market is best conceived as a set of regional markets.

For instance, according to Ive (1983), tender prices per bedspace rose in London relative to the average in Great Britain.

Regional cross elasticity of demand was high within the product market, because each of these system builders was, both technically and institutionally, in a position to supply a product which customers regarded as a close substitute for that of other firms. Client requirements were sufficiently homogeneous that each system-builder could potentially supply most of the demand of each client with their standard system. Systems were of course proprietary, and firms attempted to differentiate the product of their system, and to establish customer loyalty to their particular system, sometimes by improper methods. From an economic perspective, the corruption scandals associated with UK system building can be viewed as attempts by firms to escape from pure price competition, in the absence of other effective ways to establish client preference for their product over those of close rivals. However, perhaps more important than this, sizes of contracts were large relative to each client's total programme. For this reason most clients came to obtain most of their actual supply over several years from very few firms.

During the 1970s, this coherent regional product market fell apart. Total regional demand for the product declined, to the point where the market was no longer large enough to offer economies of scale or specialisation. This effect was redoubled by an increasing heterogeneity of project design and size. Meanwhile, on the supply side, contractors switched their methods of organisation of production, towards reliance on subcontracting to give themselves flexibility within a range of products and markets. *Economies of scope* are the savings derived from diversification, based on expert and insider knowledge of construction markets. In effect the main contractors adopted merchant contracting, providing a more managerial contracting service and increasingly hiring subcontractors to carry out the work. They moved away from reliance on specialisation in particular product-technologies, which had given them the benefits of economies of scale.

We conclude from this that only exceptionally, in the modern construction industry, will we find clear product markets, relatively isolated from other markets in construction. Nor is there a tendency towards homogeneity within product markets or a single product-market unit price. Nevertheless, the concept of price elasticity, if stripped of its assumption of product homogeneity, is a useful framework for understanding even the construction industry.

Obviously, there is a sense in which customers for construction are, in aggregate, sensitive to the relative level of construction prices compared to other prices. Clients will reduce their demand for construction in volume terms, however that might be measured, if the relative price of construction increases. The ratio between this percentage volume reduction and percentage price increase is what economists refer to as the product price elasticity of demand. Its actual measurement, however, is always notoriously difficult, and this is so *a fortiori* in construction markets.

Let us imagine that there has been a significant, but *unforeseen*, increase in the average relative price of construction. In some construction markets there are *institutional* or *behavioural* reasons for supposing that the short term price elasticity of demand of clients will be unity, so that the value of the arithmetical product of price and quantity bought is constant whatever the price level. This will be the case if clients have construction *programmes* and use certain common methods of programme budgeting. Controllers of such budgets will have no motive to under-spend them, but nor will they be allowed to over-spend. Thus, if there is an increase in the level of construction prices encountered as projects are let, the programme controllers will respond by cutting back the size of some projects or postponing the letting of projects. Total construction programme spending is thus held to its budgeted level.

In the long run, higher level decision makers within those client organisations may react to the higher construction prices relative to equipment and other prices by giving a higher priority in their total investment planning to installing new equipment in existing buildings or otherwise raising their intensity of use of buildings. This would reduce the proportion of their total investment allocated to the construction programme. If this happened then the long-run price elasticity of demand of those clients for construction would be greater than unity (i.e. elastic) Customers would be responding to the higher relative price of construction by making it a higher practical objective in their decision making to economise on the amount of construction required. If this requires, technically, greater quantities of other inputs then this will be a price more worth paying, the higher the relative price of a unit of construction compared to a unit of those other inputs.

On the other hand, construction projects are often integral parts of wider investment projects by clients. Within a total project budget, construction costs may be dwarfed by the other costs incurred including equipment investment, land, operating costs and finance costs. The short run construction price elasticity of demand of clients involved in major projects may be less than unity, and even approach zero. Redesigning a project to reduce its construction content might be an option but if it held back a whole project the additional cost of delay would usually be greater than the benefits of any savings in construction expenditure. In this case, project contingency funds will be used to cover the increased total construction expenditure resulting from higher prices.

Of course, construction prices can go down as well as up, especially during recessions. The question then is to what extent cheaper construction prices increase demand for construction. There is evidence that projects are sometimes brought forward or accelerated to take advantage of lower prices. However, mostly clients are not attracted to employ contractors on the basis of lower construction prices, but employ contractors in response to demand in other markets which require buildings or works to be erected.

In the long run, lower construction prices would not mirror the consequences of lower prices of say, consumer goods. When consumer goods go down in price, demand for them increases. If construction contracts fell in price, land prices would rise but property prices remain unaffected. The reason is that in the long run, competing developers would raise their bid prices for sites. Lower construction prices are therefore not passed on to users or future owner-occupiers but are paid to land owners who sell their sites at increased prices. As buildings (property) remain at the same price, other things being equal, there is no increase in demand for buildings (property) and no increase in demand for construction.

However, where construction prices are reduced, a certain amount of substitution may take place in the long run between the amount of construction and the intensity of use of buildings. Also likely, however, is that where construction prices have been reduced, buildings become more complex as designers attempt to spend up to budget.

Over time, in fact, construction prices in the UK and elsewhere have tended to rise substantially relative to the price of both other capital goods and consumer durables. Ball (1988) and Ball and Wood (1994B) have pointed out that the long run rise in relative construction prices mainly reflects the slower rate of labour productivity growth in construction compared to other sectors of the economy, whilst construction labour wage rates have risen in line with wage rates in general. There is some evidence, according to Ball (1994), that this has caused decision-makers to seek more building-saving solutions to their requirements. Such long run trends mean that certain relative price changes will be foreseen, by well informed decision makers. Their response to such trends will be quite different from their response to sudden, unforeseen price changes, of the sort we considered above.

Based upon extrapolation of past trends, the conventional wisdom in the UK in recent decades has been that over time the relative prices of labour, energy and construction will all tend to rise. Decisions about long term investment projects could reasonably incorporate all three extrapolated trends as assumptions, putting more effort into economising on the use of these three inputs than could be justified by their present prices. However, there is also much evidence that most decision makers focus on short payback periods or use high rates of discount, and both of these methods of appraising investment reduce the effect of such anticipated future trends.

Market segments

Recent works of business economics, for example Neale and Haslam (1994), approach the study of market and market-composition through the notion of product market segments. Unfortunately for us, this idea has been developed with branded, multiple products in mind. The analysis begins by assuming a set of firms each producing a set of branded products or product lines. Each unit of output of a line is identical, or at least very similar. Thus total output of a firm is:

$$Q = \sum_{i=l}^{i=j} q_i . L_i \tag{4.8}$$

where

Q = total output of a firm
L_i = a product line
q_i = the number of units of that line produced
j = the number of product lines

Each product, though branded or otherwise differentiated in terms of quality, is sold in competition with some fairly close substitute lines sold by other firms. (Lines are substitutes if buyers treat them as such, and choose from within that range.) The identity of these rival firms may differ from brand to brand. Any set of lines that are close substitutes constitute a *product* and a product market consists of their producers and buyers. As product markets develop through time the number of lines or product variants offered in total by all firms will tend to grow, either because each firm offers more lines, or because more firms enter the market. At this point separate market *segments* tend to develop.

Thus whereas initially, in the 1920s, a car manufacturer might have produced three product lines (say, a small car, a medium car and a large car), soon it produces each in many variants (say, 2- or 4-door; estate, saloon or hatch-back; 'popular', 'touring', 'sports'; diesel or petrol) and some of these variants succeed to the point where they become treated as new lines. At the same time whole new lines are added to meet specialist demand (e.g. 4-wheel drive, off-road vehicles).

Initially, when a total product market is relatively small, it makes sense to producers to differentiate the product only to cater to the biggest sub-sets of demand (for example, size of car). Later, as the market grows, even a highly specialised demand accounting for only a very small fraction of total demand becomes big enough to support production of specialised lines. Moreover, since customers with only 'standard' requirements already have many competing products to choose from (each firm in a market will be producing such a line), one way to get ahead of competitors is to identify specialised demands and be among the first to supply them. By this point in the history of a product market, many lines will no longer be treated by consumers as close substitutes for their chosen line, and there exists a multiplicity of markets.

Of course, such firms will not just wait for special demands to make themselves known. Instead they will try to persuade consumers, through means such as advertising, that their *new* line embodies features that they have really always needed, but not realised until now. Also, of course, they will use the same means, plus fashion styling, etc. to try to persuade buyers that each line presently on offer really is *new*.

Specialisation by construction firms

Different firms possessing capability in different techniques are required for different types of construction work. Indeed if one defines an industry by the technology, plant and skills which it uses, then the construction sector is composed of several separate industries. The building industry is distinct from the civil engineering industry. However, if we ask, with NEDC (1978), 'how flexible is construction?', the answer by definition will be that resources can be allocated flexibly between uses within an industry, but not between industries. In that case, the question becomes: how many industries comprise the construction industry, and which are they?

Some construction firms are defined by their capability in a particular resource technology, and specialise in work involving that technology. This approach could be applied, for example, to concrete work and brick work. This is the modern equivalent of the traditional craft trades, each defined primarily by the material with which they worked. For example a plumber was traditionally a worker in lead, which might include lead in windows as well as in pipes or on roofs. A mason was and still is a worker in stone, which might include walling, carving or sculptural decoration.

Other firms, such as roofing contractors, specialise in a particular building element or sub-assembly. The output of such a firm is not usually a whole final built product, since buildings and structures combine many sub-assemblies. See Ive and McGhie (1983).

Mostly, both these types of firms enter the contracting systems as subcontractors. Their market consists of buyers who are themselves construction firms.

Other firms specialise in a particular type of output to service the needs of a final product technology, such as constructing petrochemical process plants and water treatment plants. These construction firms tend to be subsidiaries of the biggest construction groups. Firms specialising in these high technology aspects of construction are mostly to be found in engineering construction rather than building. They tend to operate in international markets to protect themselves from fluctuations in local demand, and to achieve the required efficient scale. Their actual expertise may not lie in all of the input technologies required to make their product type, but only in certain key ones. Their key specialist knowledge is sufficient to give them the lead position in the supply of the whole product, with other specialists being brought in as their subcontractors. These firms are closest in the construction industry to the model found in industrial economics textbooks, of single-product firms, a set of whom constitute an identical industry and market.

Other firms practice specialisation according to customer rather than product. That is, they have a defined base of major customers, and attempt to meet enough of the construction demands of those customers to obtain their turnover targets. Traditionally, many architecture and other design firms operated in this way. Likewise, public authority DLOs were in effect an extreme case, with just one customer to serve. Also many subcontractors rely for their work on a narrow base of customers, in this case comprising main contractors. However, subcontractors are more of a hybrid case, since they also practice production specialisation. Customer specialisation saves on transaction costs and is a way of coping with demand uncertainty. However, because it does not involve production specialisation, it is unlikely, therefore, to economise on production costs. These firms face a market comprised of a set of buyers, but not for a single product type.

Other firms specialise by type of transaction or by market institution. That is, they develop expertise in one transaction-stage of a particular social system of organisation of the construction process. For instance a construction firm may consider itself to be a specialist work-package contractor working on large scale speculative office developments, under management contracting relationships.

Still other firms in building and civil engineering specialise by project characteristics, such as size or degree of complexity. Here, it is useful to think of u-shaped cost curves in terms of the relative costs of a single project. Relative costs mean the firm's costs relative to those of its effective competitors for that project. The u-shape implies that there is an optimum range of project size or degree of complexity within which the firm is at its most cost-competitive relative to rivals. If it takes projects outside those ranges of project characteristics, its relative unit costs per project increase, leading to lower tender success rates or lower margins. It is important to note that, unlike normal textbook cost curves, these are relative to costs of other firms, and relate to the project and not to the total output of the firm.

Most important of these modes of specialisation is the fact that the construction market is structured first into the contracting system and

Table 4.2   **Matrix of markets defined by specialisation**

	Type of firm		
	Engineering firms	**Roofing contractors**	**Speculative housebuilder**
Customer	Narrow specialised oil extractors	Main contractors	General public
Final product technology	Off-shore oil platforms	Large span commercial building	Dwellings
Transaction type	Negotiation with clients	Tendering and negotiation with contractors	Marketing and selling of completed houses
Project size and complexity	Large with emphasis on quality.	Medium sized, inc. refurbishment	Small scale batch production

the speculative system, and then by its division into different versions of the *contracting system*. These contracting systems to a large degree correspond to types of work. Wherever such a correspondence exists, then there is a highly distinct sub-market, as well as a sub-industry.

We have defined contract construction systems (Chapter 2, p. 30), according to the *system* of relationships adopted between actors and the organisation of their respective roles. These systems, which can be used to identify markets from the point of view of firms, are therefore distinct from but overlap procurement routes, which are sometimes over-narrowly defined in terms of contract forms. Each system, furthermore, has its own characteristic forms of competition.

Specialisation in housing production

This approach to specialisation by construction firms can best be understood by applying it: for example, to the production of housing. The particular specialism of a housebuilder *may* be the technology of the final product, namely dwellings. However, the technology used for house building in the UK is so widely diffused that it is unlikely that any one firm could have a specialist advantage in it.

Alternatively, a firm might possess production advantages from learning-by-doing, or from economies of scale in production. However, because of the high level of subcontracting, housing production is rarely organised in ways that would permit a house building firm to take full advantage of production techniques learned on one site in subsequent jobs. At best this puts the firm on a par with other experienced house builders, and at an advantage only over firms without such experience.

If a housebuilder specialised in the technology, we would expect the firm to operate whenever the product technology were of a particular type, across both speculative and contract housing markets. This is relatively rare. Nor do housebuilders specialise in terms of a customer base. Houses are mostly bought by private households and, by definition, these are only occasional customers.

If housebuilders' specialisation were based on transaction type, they might, for example, specialise in selling houses to one-off, inexpert, purchasers not employing professional advisors. These buyers would be willing to purchase 'off-the-shelf', using specific forms of house purchase finance, for whom mass marketing techniques could be applied. If this type of specialisation were the case, the same firms might be expected not only to build houses but also produce other fixed *in-situ* consumer durables, such as fitted kitchens or double-glazing or conservatories. These firms would also be in a position to transfer expertise and personnel from house building to these other products.

Finally, housebuilders might specialise in land acquisition, land development and sale. In this case, there might well be some overlap with other kinds of land development activity.

However the particular specialisation of housebuilders is, in effect, based on a combination of a specific type of product, namely housing, and a certain type of customer, namely households, together with the social *system* of organisation of the construction process (SSOCP) found in speculative building, which involves the integration of the roles of land developer, housing developer, designer and construction manager, but not final owner.

Concluding remarks

In this chapter we have seen that there are institutional market factors, for example, the minimum efficient scale of operation, which arise out of technical factors, such as the size and operational requirements of the plant used. The minimum efficient scale of production thus contributes to the degree of seller concentration in any given market. However, the actual behaviour of firms is based more on a search for modes of specialisation in market segments, which in turn are defined in complex ways which are partly based in technologies but mainly in SSOCPs.

Actual behaviour in construction markets by construction firms can be described using a neo-classical framework but the assumptions of a neo-classical approach need to be modified to fit a description of what actually occurs in construction.

5
Construction Markets and Transaction Costs

Introduction

This chapter discusses the way all the construction activities that go into the production of a finished product are co-ordinated. Co-ordination may be achieved either by command or by market exchange. If all activities are carried out by workers directly employed by one firm, co-ordination is achieved by command within an hierarchy. The hierarchical power of employers enables them to give orders to all their different employees, who therefore may be envisaged as the many *hands* of a single controlling *brain*.

Alternatively, each direct producer (worker) could be a small independent self-employed producer trading with other producers, either directly or through the mediation of capitalist merchants. In this case, co-ordination is achieved entirely through the 'invisible hand' of the market. Each producer acts independently of all the others, but each is motivated by self-interest to respond to the same price signals generated by market exchange.

Co-ordination of the construction process can also be achieved through a third alternative in which the stages of production are divided between several capitalist firms, each employing a work force, and each trading with the other firms. In this case, which is the normal one in our society, co-ordination is achieved by a combination of command and market.

In particular, this chapter looks at transaction costs which are usually present in one form or another whenever a transaction (trading) occurs between two separate parties. These transaction costs are often hidden but have their origin in market friction caused by ignorance, lack of resources or uncertainty. In this chapter we investigate how these transaction costs impact on the cost of construction, and on the organisation of the construction sector.

Markets and hierarchies as means of co-ordination

Co-ordination of the construction process has two aspects: one concerns relationships between workers and capitalists; the other, relationships between capitalist firms. Capitalists will normally prefer to organise their relationships with workers as ones of command rather than exchange. They would rather employ their workers than buy the workers' output from them. This is the difference between an industrial firm and a mere merchant. We discussed this earlier, in Chapter 1, in terms of the difference between formal and real capitalist control over the labour process (see also Ive and Gruneberg, 2000, ch. 2). This is partly a matter of power. Because employers are generally able to exercise power over their workers, they are normally able to extract from their labour an output with a value in excess of the wage they need to pay. There is no automatic tendency in the labour market for the value of the wage to rise to equal the value of a worker's output.

If capitalists were to buy the output of independent producers (as was often actually the case in the early history of capitalism), they would only be able to make profits, from reselling that product at a higher price, if there was some imperfection in the market so that different prices could co-exist for the same product. In early capitalism, with very poorly developed transport and communications, and with each local market barely in touch with other markets, it was indeed easy for merchants to make such *commercial profits*, from buying cheap and reselling dear in geographically separated markets. Today, though it is by no means unknown, it is harder for capitalists to make commercial profits from geographical barriers. However, as we shall see, there are other kinds of barrier that may serve to keep producers separated from direct trade with the ultimate customers for their products, and in these cases commercial profit is still possible. Capitalists do not have to take direct control of production in order to make profits.

Relationships between firms in a production process can vary from complete integration of all stages under a single firm to a high degree of specialisation by firms in a single part of the process and a consequent high degree of fragmentation. It is helpful to think about the inter-relationship of firms by first imagining two, artificial, extremes to the solution of the problem of economic organisation. The first is an entirely market structured economy, while its opposite is one which is entirely command structured or hierarchical. All actual forms of economy can then be placed on a continuum between these poles.

In a market-only economy each individual producer would be, as it were, a one person firm. All their economic relationships with other people would take the form of exchange relationships, as between buyer and seller, and these exchange transactions would be governed by market forces. Then, the study of the economy would amount to a study of markets, the formation of prices, and of the responses of individuals to market price signals. There would be nothing else there to study.

In a command-only economy there would be just one giant firm. All production would be done within departments of this firm, and all decisions as to what to produce and how to produce it would be planning decisions by the management of this firm. All producers would be employed by this firm, and would work to its commands. There would be no markets. Instead the firm would simply issue orders about the distribution and use of all inputs and outputs. With only one employer, there would not even be a labour market. Workers would simply be allocated part of the output belonging to the single *firm* for their consumption. In such a scenario, the study of the economy would amount to a study of plans and the decision making of the planners of this single integrated production system.

Though exaggerated, these two pictures correspond respectively in some degree to the images of capitalist and communist economies to be found in economic ideology. Our present concern is solely with capitalist economies. The notion that modern capitalism is essentially best seen as a series of markets connecting individuals has been very influential. This view of modern capitalism has emerged despite the fact that it completely fails to capture one of the most important aspects of capitalism as it really is, namely, that most production is actually organised by large firms.

Adam Smith, the founder of classical economics, asked his readers to imagine an 'early and rude' (anti-) society of independent individual producers (specifically, of hunters) drawn ineluctably into trade with one another by the advantages of specialisation and division of labour (see Skinner and Wilson, 1975). Imagine, Smith wrote, a tribe of hunters, in which one person specialises in the production of bows and arrows whilst others hunt the animals, some in turn specialising in hunting deer and others beaver, because of their greater dexterity at the one task. By exchanging bows for carcasses and beaver for deer, each party ends up with more carcasses (consumer goods) than they would have had without that division of labour (Smith, p. 19). Smith considered exchange to be the main means whereby the division of labour could arise and be developed. In fact in the same work Smith discusses the division of labour within a capitalist workshop and firm (the famous pin factory), but the ideological pull of his account of the essence of society (of social relationships and of the economic system) as lying in the decisions of pre-existing individuals to trade their products with one another, rather than in the relationship of an employer with their workers, was overwhelmingly strong both for a petty bourgeois readership (shopkeepers and other small businessmen), for this was how they saw their own world, and for intellectuals brought up in the British tradition of 'possessive individualism' and the social philosophy of John Locke. A century later Walras and Jevons, the founders of neo-classical economics, formalised into a mathematical model this idealisation of a perfectly fragmented, perfectly competitive society of independent individuals.

Transaction costs

Orthodox economics has always struggled to come to terms with the actual reality of an economy dominated by a relatively small number of very large firms. Recently, transaction cost economics has been developed to explain the existence of firms, which operate in a way which is contrary to the founding postulates of competitive market economics.

For the sake of simplicity, let us define a *firm* as any unit which organises production. The transaction cost approach explains why these units evolve beyond the size of the single individual, and why they grow as big as they do, but no bigger.

Every transaction between two parties involves hidden expenses called transaction costs. These hidden expenses create a cost to every transaction but are rarely taken into account either by accountants or by economists. This is because both economists and accountants have traditionally associated costs with production. It is also because transaction costs are often hard to identify and are buried within the general costs of the business, or else appear as inefficiencies, reducing revenue or increasing cost beyond what it would otherwise have been. However, certain transaction costs can be reduced if firms choose to produce the product or service in-house instead of going outside to procure it. On the other hand there may be reasons why a firm might wish to shed direct responsibility for a certain production activity, which it presently undertakes. In this case it can use market exchange with other firms to buy-in what it needs rather than make it itself. If other firms can organise this production activity more efficiently than could the original firm, production costs will be lower. But both buyer and seller will then face transaction costs. Total cost will be reduced if the sum of production and transaction costs is now lower than the previous level of production cost.

For economic orthodoxy, in the tradition of Smith and Walras, the very existence of firms is a problem to be explained, whereas the independent producer appears as the *natural* state of affairs. Why, asks Williamson (1975), don't the workers in a factory each operate as one person businesses, selling their output to the worker on the next work station down the line, buying their input from the one upstream of them, and each renting their share of any jointly used space and equipment? His answer is that it must be because the costs of organising all the resulting transactions would be too great – greater, that is, than the costs of organising relations of command through employment contracts and wage labour transactions.

It is not that transactions disappear with the use of the employment relationship. They are only reduced to a limited and occasional negotiation of the terms on which wage labour will be sold by the worker and bought by the employer. The typical employment contract makes full allowance for unforeseen contingencies. The worker, during the period of the working day, will perform within certain broad limits whatever tasks they are directed to do by the representative of the employer. Thus there is no need for a separate contract to be negotiated each time the content of the work to be done changes, whereas in exchange relationships between independent producers, new contracts would be needed each time.

Beyond this, there is the question of the number of activities or processes which a firm performs *internally*. In the transaction cost approach, it is assumed that the scale at which any one activity is carried out is determined by the economies of scale of production (see below). The size of a production unit depends on this, and on the number of linked activities it groups together within itself. This number is explained by transaction cost analysis.

According to Williamson transaction costs are caused by *bounded rationality* and uncertainty. Bounded rationality refers to the fact that firms do not possess all possible information relevant to their decision making. Because of the cost and delay involved in trying to obtain complete information, firms choose, or are forced, to act on incomplete data. Their decisions then, though *rational* in the sense that they are based on calculation of the best means to achieve certain ends, are only *boundedly* rational because they are not based upon full information.

Whereas information about the past is in principle available, though at a cost, the future is unknowable. It is not that firms have incomplete *knowledge* about the future. By definition, there can be no such thing as knowledge about an *uncertain* future in an economic world subject to continuous change, both from within and without. It is not even possible to attach accurate probabilities to future events or occurrences in most cases. If it were, we would speak of (statistical) risk rather than of pure uncertainty. The more complex a transaction is, the greater the uncertainty, and the more serious become the implications of bounded rationality.

Information impactedness, or the unequal possession of information or knowledge on the part of the firms party to a transaction, is also a factor in raising transaction costs. The cost of acquiring information is a major hidden cost and a compromise must be reached between the cost of acting with incomplete information and the cost of obtaining complete knowledge about a product or service. Information impactedness makes opportunism a problem.

Opportunism, the willingness of transactors to take advantage of a situation at the expense of their co-transactor, specifically, to conceal information or to make self-disbelieved statements or promises, is a problem wherever there is information impactedness, and wherever it is costly or impossible to use complete contracts (see below). Only with complete contracts is a promise, from a possibly opportunistic transactor, as good as performance. 'Opportunism' is a more serious problem if the single transaction is large and unlikely to be repeated (since the main resource of a victim of such behaviour is to exclude the opportunistic transactor from their future transactions), and if reputations of transactors are relatively unimportant to them. It is a less serious problem if the transaction is small, repeated, and for a homogeneous product, with many competing buyers/sellers.

Each party is assumed to be self-seeking (and, potentially, 'self-seeking with guile'), and therefore prepared to take advantage of the other if they can get away with it.

In the presence of endemic uncertainty contracts are devices used to try to anticipate various contingencies or changed circumstances, and to lay down how they will affect on-going transactions. However contracts are expensive to devise and negotiate, and even more expensive to enforce in the courts. Moreover it is impossible for the framers of any contract to foresee all possible contingencies, especially where, as in construction, transactions are of long duration.

Power in transactions lies with those with the best information about the product to be exchanged or the service to be provided as well as any alternatives and their prices. Advantage also rests with those in positions of relative monopoly rather than those with many direct and obvious competitors.

Transaction costs in construction

The market mechanism is used in the UK construction sector as an alternative to the management of the production process by a unified management team. In UK construction, project design is often divided between many firms. The building process is invariably carried out by a number of firms on site, while building components are manufactured by other firms off site. The construction industry can therefore be characterised as *market orientated* rather than hierarchy orientated. Each time an intermediate product is passed on or an intermediate service is carried out during construction, it is necessary for the firms to engage in a transaction. Each transaction negotiation reinforces the separateness of firms in the construction process.

According to Winch (1989), managers in the UK construction industry, responsible for the same project but on behalf of different firms, are thrown into conflict with one another as they represent their own firm's best interests and as these interests tend to diverge. The confrontational characteristics of the construction sector and the transaction costs incurred need to be understood within an institutional framework based on traditional practice and legal constraints and liabilities.

Markets economise on production costs but not on transaction costs. For this reason it is possible for quantity surveyors to demonstrate that construction costs have been kept to a minimum by accepting the lowest priced bids, whilst at the same time the industry is accused of inefficiency. This apparent paradox can be explained by examining transaction costs which arise during the sequence of actions involved in many construction projects.

Below, we use two frameworks, an empirical approach and a theoretical approach, to examine the factors determining the level of transaction costs in construction. The first framework is empirical, and follows the sequence of actions involved in a construction transaction. The second is theoretical, and follows the conceptual categories identified in the transaction cost economics literature.

An empirical approach to transaction costs in the construction industry

The measurement of transaction costs is difficult. However, for a construction buyer they certainly include all of the following.

- *Search costs* are the costs of finding out information about who is offering what products or services and at what prices. In construction, this means identifying who might be the best or most competitively priced suppliers for the particular project, from amongst a vast population of firms.
- *Product or service specification costs* arise because a market supplier will only provide what they have contracted to supply. The product or service to be supplied therefore has to be specified very carefully and fully in order to obtain the desired quantity and quality. In some other industries the product already exists before it is bought, and can be examined. In construction, this is normally not the case. Examination must therefore give way to specification, which is a more expensive activity.
- *Contract selection, contract design, and negotiation costs* are the costs of finding or creating forms and conditions of contract that are suitable to the particular needs of the buyer in the particular transaction in question. These terms then have to be agreed with potential suppliers, who will of course have their own particular needs and contract preferences. The agreed terms usually emerge through a process of negotiation and compromise, though some buyers may sometimes be in a strong enough position to impose their contract terms unilaterally. Contracts specify both 'what is supposed to happen' and 'what if the unexpected happens or what is promised does not happen?' Contract forms and conditions deal largely with contingencies and promises – what will be the rights and duties of the two parties *if* such-and-such happens in the outside world, and *if* one of the parties defaults on one of their contracted promises. What will constitute a default, and how will it be identified? If a dispute arises, who will adjudicate upon it, and how? Because the circumstances of both construction projects and buyers vary so much, it is often not appropriate or possible to use a standard pre-existing form of contract, and this of course increases this category of transaction costs.

- *Supplier selection costs* occur if the product or service to be purchased does not have a single, uniform market price, as is the case in construction. Normally a price competition will have to be organised between potential suppliers, with the contract awarded to the lowest priced tenderer. If potential suppliers differ in the quality of what they offer, selection may involve making trade-offs between offers of different combinations of price and quality. Winch (1989) points out that contractors subcontract to specialist firms who in turn may subcontract to others. Each time work packages are put out to tender several firms are asked to estimate their costs and bid for the work. Generally only one firm is selected. The cost of operating this market mechanism includes the costs of preparing failed tenders, which each contractor then passes on to the client in the form of prices which include the transaction cost of the tendering process.

- *Contract performance monitoring costs* are transaction costs incurred due to the need to measure and control performance in terms of its price, timing and quality. In construction it is not sufficient to measure a supplier's performance once only, on completion of the contract. This is because the costs of remedying poor quality become exorbitant, whilst quality itself becomes very hard to measure, once the work in question has been incorporated into a finished building. Time performance obviously must always be continually monitored rather than just measured after the event if there is to be scope for corrective measures. Surveillance and control systems add to transaction costs through the need for scheduling models, continuous cost accounting and performance measurement.

- *Contract enforcement costs* are the cost of legal bills and delays. These costs may arise if one party thinks that the other has breached its contractual promises and the terms of the contract can be enforced, in the ways laid down in the contract and implied in general law. Some contracts contain arbitration clauses, since non-judicial dispute resolution is normally cheaper, but parties can reserve the right to reject arbitration and insist on taking the matter to court. If the supplier will not or cannot meet its promises the contract may allow imposition of financial penalties, called liquidated and ascertained damages, to compensate the buyer for non-performance, but this compensation does not necessarily prevent delay. Any delays to the completion of projects add interest charges to developers' financing costs. Even the threat of legal action, if it is to be credible, involves high costs of preparing evidence to support the claim.

The factors which determine whether or not markets are a reasonably efficient mechanism for co-ordinating relationships between the producers involved in each process necessary for the production of a finished product are: the frequency of, and uncertainty associated with, transactions. The more frequent a transaction, and the greater the uncertainty involved in a transaction, the more likely it is that the cost minimising solution is to bring both activities within a single firm. If transactions are occasional and involve relatively little uncertainty, a firm will normally find that the cost minimising solution is to make a set of contractual arrangements by contracting out work to legally independent and separate firms or self-employed individuals. In this case the firm chooses to buy in, rather than make, products using markets for subcontractors. If transactions are infrequent, but highly uncertain, the firm must consider both production and transaction costs carefully. The uncertainty argues for internalisation but, if the requirement is infrequent, internal production may be relatively expensive in terms of production costs. Special solutions may be called for.

An advantage of hierarchies over markets is that continuity of work within a firm means that experience gathered on one project can often be applied to the next. In much production, it is found that any organised set of producers improve in their efficiency as they learn by repeating the same technical tasks in the same social context (Bishop, 1975; Jackson, 1982). This point rather breaks down the neat independence of production and transaction costs. Dietrich (1994) expands upon other possible sources of interdependence of the two in a dynamic context.

A problem of markets is that they do not provide a feedback mechanism to members of separate firms. This is an aspect of information impactedness. Feedback of information across boundaries between firms is impeded and less perfect than feedback within one firm. Moreover, feedback is limited because market relationships imply impermanence, if they are governed by incessant competitive tendering. That is, project *teams* (comprising people from a set of firms) may be broken up by different firms winning the next supply contracts or subcontracts. The greater the importance of feedback and the complexity of the system within which learning curves apply, the greater the advantages of integration of a permanent work force within a single firm. Hierarchies also reduce the costs of coping with risk and uncertainty and increase control by management over the production process.

On the other hand, hierarchies mean that firms can no longer rely on *market disciplines* to achieve efficiency in the use of resources. Market competition, by ensuring that supply is always from the cheapest source, is supposed to ensure that it is also always from the most efficient source. In fact *technical efficiency*, in the sense of using inputs with the least total opportunity cost, may not coincide with *lowest market price*, because market prices of inputs to production are not always good indicators of the social opportunity cost. However, leaving that objection aside for the moment, it is clear that once a firm buys-in resources from which to make a product, rather than buying-in the product itself, it has to find some non-market means of ensuring that it is using the best combination of resources and using them as efficiently as possible or, at least, as efficiently as its nearest competitor. It is also clear that this may be a rather hit-and-miss process. If the firm does this poorly for all its activities, this will soon become evident, as the firm's total unit costs for its finished product will become uncompetitive with those of rivals. However, inefficiency in one department or branch could well be hidden within an overall picture of broadly competitive costs.

A theoretical approach to transaction costs in the construction industry

It is possible to apply Williamson's general theoretical concepts of transaction costs to the construction industry. Several factors generate transaction costs in the construction industry.

- *Uncertainty* in all construction projects is a major source of transaction costs. The complexity and uncertainty associated with construction projects is combined with *bounded rationality*. The sources of uncertainty in construction stem from one-off production, the duration of each project and the uniqueness of each building scheme. As a result it is not always possible to know, in advance, precisely what tasks will need to be undertaken on any project. Weather and ground conditions also lead to uncertainty and add to the risks of undertaking construction work. There are further uncertainties due to the temporary nature of the organisation of construction projects. For instance, project teams may be made up of people who have not worked together before and whose interaction is therefore uncertain. Also, *duration* creates uncertainty, as it increases the possibility of new circumstances arising that could change the value or cost of an original plan, and thus generate a desire to deviate from that plan.

- *Small numbers* in construction relates to the one-off nature of building projects. *Opportunism* means that suppliers can take advantage of the difficulty in pricing work when only a small number of competing firms are involved. Once negotiations are started the number of potential suppliers facing the buyer drops further, and once the contract is signed the contractor is in a position of temporary near monopoly (Hillebrandt, 1984). The client can terminate the contract if the contractor tries to take excessive advantage of this situation, but only at a great cost. Clients also have scope for opportunism, because of the small number of buyers facing some firms.

- *Asset specificity* in some industries refers mainly to plant and machinery and to the necessity of employing a particular specialised piece of equipment for a given task or function. In construction it is usual for plant hire firms to be able to provide all necessary machinery. The low asset specificity of construction plant means that there is relatively little incentive to establish permanent relationships between the owners of specific plant assets and the customers requiring use of those assets. In construction, asset specificity refers mainly to investments in information systems and in human resources. For example, if staff of one transactor acquire implicit knowledge of the idiosyncratic needs, requirements and procedures of another, this knowledge loses its value if the exchange relationship with that transactor is ended. If asset specificity is high, that is if each *set* of projects for a client are most efficiently performed using assets developed specifically for that purpose, two things follow. First, those firms which already own such assets will be at an advantage in tendering over those that do not, in that they will not have to include the full cost of acquiring such assets in their estimate of project cost. Secondly, clients will be able to exert opportunistic pressure on those suppliers who have already committed themselves (on the basis of client promises of a series of contracts) to investing in assets with limited value in alternative uses, forcing them to cut their profit mark-ups in subsequent tenders, as the opportunity cost to them of not winning the contract would be substantial.

Each party stands to gain from supplier investment in specific assets, but the supplier will not do so unless given some guarantees of continuing use for them. The result is likely to be either integration of the two firms, or else some long-term commitment, going beyond the single project, to use the specific supplier.

- *Information impactedness* arises from the withholding of information, which adds to transaction costs as transactors are deprived of technical or market information ideally needed for accurate pricing, decision making and progress monitoring.

An alternative approach to the 'problem' of the large firm

Instead of taking the large firm as an anomaly to be explained, suppose instead that we begin by recognising that the firm, equally with the market, is one of the defining institutions of capitalism. Firms exist and, like any institution, tend to perpetuate themselves. They make profits, and prefer to reinvest these profits within the firm, rather than elsewhere. In this way firms accumulate capital by retaining profits. To see how firms achieve this, let the average rate of profit on capital employed, after tax and interest but before dividends, and over the long run, be x per cent, and let us call this the net profit rate of the firm. Profit remaining after dividends have been distributed to shareholders is called retained profit. Let the average retention ratio be y per cent.

Hence

$$y\% = \left(\frac{\text{Net profit} - \text{Dividend}}{\text{Net profit}} \right) \times 100 \tag{5.1}$$

Retained profit as a percentage of capital employed, which we shall call z, will then be $(x.y)$ per cent. Thus, supposing that firms by and large follow Polonius' injunction ('neither a borrower nor a lender be' – *Hamlet*, Act 1, scene 3), their capital employed will accumulate at a rate of z per cent per annum. Suppose their initial capital employed is K_0. The mass of retained profit and capital accumulated within the firm each year, Z, is given by $(K.z)$. Thus

$$K_0 + Z = K_1 \tag{5.2}$$

An example will help to illustrate the relationships. Assume capital employed is initially £100,000. Assume the annual rate of net profit is 10 per cent of the capital employed. Net profit is therefore £10,000. Of this profit, assume 20 per cent, or £2,000, is retained. Then Z is 20 per cent of the 10 per cent profit, £2,000, and z is 2 per cent of the capital employed. If the accumulated capital of the firm were £100,000 at the beginning of the year, K_0, then K_i, or $K_0 + ZK_i$ would be £102,000 by the end of the year. The rate at which the firm expands its capital is Z/K_0.

Bear in mind that firms can expand their assets at a faster rate than Z/K_0 if they borrow in order to invest. In this case $K_0 + M$ is their initial assets, where M is their initial borrowing. Total retained profits available to be added to capital employed next year will be Z plus W, where W is the excess (or deficit, if W is negative) of *gross* profits made from using M over the interest cost, (i), of borrowing M, $M.i$.

In order to compute the rate of capital accumulation for a firm using borrowing, we need additionally to know the gross rate of return on assets (before deduction of interest costs from profits, and before deduction of debt from assets). Let us call this rate u. Finally, let d be the 'cost of equity capital', the dividend paid-out expressed as a percentage rate of return on shareholders' funds, K.

Now $W = M.u - M.i = M. (u - i)$
Let us represent the ratio W/M by w.
The capital accumulation rate is now:

$$\frac{w.M + z.K}{M + K} \quad \text{or} \quad \frac{W + Z}{M + K}$$

or

$$\frac{M (u - i) + K.(u - d)}{M + K}. \tag{5.3}$$

Now, how is the firm going to use this additional capital? Several strategies are possible. For instance, the firm could expand the volume of output of its present activities, using its current technology and methods. Alternatively, it could increase its capital:output ratio by switching to a more *capital intensive* method of production, or it could add on some new additional activities. Let us assume, as may normally be the case, that $z \geq n$, where n is the rate of growth of output in the construction sector at large, and therefore the average rate of growth of the market for a typical product. If z is greater than n, then the firm has three options. It must either try to increase its share of the total market for its present activity, or diversify into other activities, or become more capital intensive in its methods.

Taking the last option first, a firm would choose to increase the capital intensity of its production methods in order to reduce total cost per unit. If we suppose the firm is already using the least cost method of production, then we can rule out this option. In fact over time technical change makes more capital intensive methods profitable, and this in itself favours those firms able to command larger amounts of capital.

Assuming that firms already use the least cost method of production, then the average firm is left trying to increase its total output at a rate faster than the whole sector is growing. If an individual firm succeeds in this, then necessarily it becomes, both absolutely and relatively, larger as a firm. However, if the total number of firms remains the same, by definition they cannot all succeed in outgrowing the sector at large, and some must shrink in order for others to grow. Alternatively, the total number of firms in the sector can fall. Competition between firms is then the process which permits some to grow at the expense of others, which shrink or cease to exist.

Moreover, there is evidence, according to Mueller (1986; 1990), that the relative success of a firm in one time period is positively correlated with relative success in the next time period, and so on. It is therefore possible to describe a system of cumulative causation, of virtuous and vicious cycles, in which 'whosoever hath, to him shall be given, and he shall have more abundance: but whosoever hath not, from him shall be taken away even that he hath' (*St Matthew's Gospel*, chapter 13, verse 12). Without this tendency to reinforce success or reinforce failure, though firms' accumulation and profit rates would differ from one to another in any one year, over time they would tend to even out.

This process of cumulative causation occurs in the context of firms competing with one another in an uncertain and continually changing economic context. Adaptation to these changes normally requires investment of capital, whether to change methods of production or to change products. Firms that have recently been unsuccessful will lack the retained profit to make these adaptive investments, and are therefore likely to become still less successful as time passes. It is also possible that there are important economies of scale, either to scale of a production unit or of an ownership unit. An ownership unit may be a parent company which owns several completely separate production units. Large ownership units are often in a position to take advantage of financial economies of scale. These parent companies are usually able to raise additional capital, beyond their retained profits, more cheaply than smaller ones, either by borrowing or by issuing new share capital. Above-average growth makes a firm larger than its rivals, if all started at the same size, and therefore gives it the benefits, if any, of economies of scale.

Unequal exchange

Given that all the above is the case, we are actually able to turn around the question that troubles orthodox neo-classical economics, so that it becomes 'why don't large firms take over all the economy, and grow without limit'? To answer this, we will explore the idea that there may be certain advantages for large firms in having alongside themselves a small-firm sector of the economy, with which they can trade.

Small firms in construction exist in two main contexts. Firstly, they operate as the main producers for final product markets that large firms do not choose to enter. The argument is that the characteristics of the market demand or the production process make it unattractive to large firms. That is, the market may be stagnant or declining in demand terms, and it may only offer relatively small economies of scale and slow technological change. It therefore offers little opportunity for sustained accumulation at a high rate, necessary to create or attract large firms.

In construction, small final producers are mainly found in the housing repair and maintenance market, although this is not an exceptionally slow growing market in demand terms, by construction industry standards. The explanation for the existence of small firms is therefore more likely to be found on the production side.

Secondly, small firms act as suppliers of, or subcontractors to, large firms. The argument is that the large firms actually find it more profitable to trade with less powerful suppliers than they would if they integrated more production activities internally. Because small and specialised subcontractors have no direct means of access to clients, they depend entirely on main contractors for their work. There are, of course, many more subcontractors than main contractors, and this alone would create some disparity of bargaining power. However, it is open to a main contractor to develop a condition of dependency in at least some of their subcontractors. This can be done by offering to *solve* the latter's marketing problem by offering them most of the work they need, by repeatedly re-using the same subcontractors on many of the main contractor's projects.

Once a subcontractor has come to rely on one main contractor for a substantial slice of all its work, it can then be said to be in a position of dependency. The main contractor can plausibly threaten to switch to another source of supply. If carried out, this would have dire consequences for the subcontractor, which might require substantial time to replace a particular main contractor as a source of orders. The buyer is then in a position more or less unilaterally to lay down the price and contract conditions. In addition to the stick, they can also hold out the carrot of more orders in the future if the subcontractor accepts the terms offered.

Firms prefer to maintain a positive cash flow, which occurs when cash entering the accounts is greater than the outflows of payments in any one period. If outflows exceed receipts to the firm, cash flow is said to be negative. During projects the cash flow may become negative as work proceeds in advance of payments from clients. However, main contractors will often be able to create a positive cash flow by delaying payments to subcontractors while receiving payments from clients. By withholding payments to subcontractors main contractors and management contractors can use the cash they hold to generate interest (Punwani, 1997). For this reason also large contractors have an incentive to permit small firms to survive.

Economies of scale and of integration

Production of any finished product has been discussed above as a sequence of activities, which may or may not occur under a unified ownership and control. We now have to address the question of the scale or volume at which each of these activities is carried on.

If an activity is not technologically divisible into separate sub-activities without raising production costs, we say that it shows economies of integration. For instance, the technology of the assembly line means that total production costs are lower if the sequence of assembly operations are all performed in the same workplace than if they are separated.

If unit production costs for an activity fall as scale is increased we say the activity shows activity (or process) economies of scale. If they rise after a point if that activity is carried out at ever larger scales we say that it shows activity diseconomies of scale.

The problem is to find the minimum scale at which it is efficient to perform any one sub-activity. This will normally differ widely from one activity to another. But, if there are also economies of integration, there will be a tendency for all activities in a sequence to be performed at a level of output required for overall technical efficiency.

For instance, imagine a factory in which many activities are linked by an assembly line, and suppose that the optimum efficient size for this assembly line is a large one, say sufficient to produce X units of final product per period. The need to balance the volume of output of each sequentially integrated activity at X means that the total output of each activity will be balanced at this same scale even though for many of them considered separately the optimum scale would be smaller. These activities can then be divided into parallel work units (for example, several workshops or machines each performing the same activity), each of the optimum scale for that activity, say Y. If Y is one-nth the size of X, then there will be n parallel units for that activity. The total size of the factory is then given by, on the one hand, the largest of the scales necessary for efficiency of an important activity, and on the other the economies of integrating those activities. The idea is that parallelism prevents diseconomies of scale in some activities offsetting economies of scale in others. So long as an activity is divisible into multiple parallel units, diseconomies of scale can be avoided. We shall see below how this applies to the organisation of production on construction sites.

It is clear that in most production activities in a modern economy there are real production economies of scale, up to a certain size of production unit. Larger production units have lower production costs per unit than do smaller ones. Why? There are two main types of reason, which may also occur at the same time.

One reason is to do with *technical disproportions* between the materials used and the capacity of an item of fixed capital equipment. For example, the cost of making a container depends more upon the surface area, which increases with the square of its dimensions, whilst output from the container depends upon its volume, which increases as the cube of its dimensions. Hence, larger vessels or containers economise on the material required to form them per unit of capacity.

The second reason is to do with *indivisibilities* of certain inputs, and consequent spare capacity or under-utilisation of those inputs at sub-optimal scales. For example, the smallest vehicle requires one driver. Indivisibility means that it is not possible to have half a driver. Increasing the size of the vehicle does not require a second driver.

Furthermore, some indivisible inputs are only needed occasionally rather than continuously, if production is carried on at a small scale. A larger office, for instance, with more staff, would not require a proportionately bigger photocopier, since most of the time photocopying equipment remains unused, and since loads upon them generated by each worker do not all coincide in time. Instead at most it requires some form of planning and queueing to spread usage through time. To give another example, if production must be continuous, maintenance staff must be permanently available to deal with any breakdowns immediately. In a small factory, most of the time there will be nothing for the single maintenance engineer to do. Nevertheless it must have one, whereas a factory ten times larger may find that the need, according to the law of averages, is for only two or three such engineers.

The *law of averages* works to smooth out fluctuations between peak and trough loads, so that the larger the production unit, the more likely it is that actual loads made up of the sum of independently varying demands will approach the average. The probability of a tossed coin coming down heads is 50 per cent. If we toss a coin 100 times it is almost certain that the actual number of heads will be within the range 45 to 55, i.e. close to the predicted number of 50. If we toss a coin just twice, however, it is quite likely that we will get either no heads or 2 heads, rather than the predicted 1.

Now, a situation that has a 25 per cent chance of occurring is not an eventuality that a firm can ignore in its planning, whereas an outcome with only a 1 per cent chance of arising will often be dismissed. Assume the number of times a coin is tossed is equivalent to the number of events occurring in a firm, such that each event has an equal chance of occurring or not occurring. If each head represents a demand on capacity (whereas 'tails' represent no demand), then our small firm would have to plan to be able to meet the 25 per cent probability of a demand as large as 2 heads, i.e. twice the average demand, whereas our large firm could safely plan to meet no more than a demand of 55 heads, only ten per cent more than the average demand. It will therefore have to carry less capacity that is frequently unused, and this gives it a cost advantage over the smaller firm. This same principle applies to the predictability and variation about the predicted mean of the aggregate demand for the firm's product from its customers. It also applies to demands which any one department or activity within the firm must meet generated by other departments and activities.

The firm reaping full economies of scale then becomes the one most likely to survive and prosper, under conditions of competition. These economies of scale are often put forward as *the explanation* for the existence of large firms. But it is also clear that the minimum efficient scale of production varies enormously from one activity to another, and that there are not always strong economies of integration of these activities. Concrete production, for example, involves a series of processes, from cement making through to batching plants and mobile mixers, each of which may have a different minimum efficient scale. On the other hand the costs of transporting unfinished cement or concrete are also high. In this case several batching plants may be served by one cement works. Transport costs as a proportion of the value of a unit of output are critical in determining the location of a firm and the degree of integration of a production process. If transport costs are higher for moving cement than concrete, all production may be grouped together at the site of cement manufacture. If cement transportation costs are relatively cheap, the concrete will be produced at or near to the location of its use in a less integrated process. These are technically potentially separable processes – that is, they do not have to be performed in the same establishment or under the control of the same firm.

Contracting-out normally reduces certainty or security of supply compared with holding in-house capacity. No subcontractor is likely to regard disruption caused to one of their customers from interruption in their supply as quite as serious a problem as the customer themselves may. A subcontractor will not be as prepared to incur exceptional costs and hold expensive spare capacity to avoid the risk of this eventuality. If production does not have to be continuous, or if it is not a problem to hold large stocks, as buffers to protect one activity from the effect of disruption to preceding activities in the sequence, processes can then often be divided into a surprising number of technically separable activities. At most, economies of scale, other than economies of integration, then define the minimum technical size of a production unit undertaking a single process.

Transaction costs and business strategy in construction

Anderson and Gatignon (1986) use transaction cost analysis to develop a strategy for firms entering new markets or embarking on new projects. They begin by assuming that, in the absence of information to the contrary, a low level of ownership of specialised plant and equipment is desirable in order to minimise risk. They label this premise the 'default' hypothesis. This follows from the fact that such assets are to some degree *specific* in the uses to which they can be put. They have been designed for certain purposes, and will be less effective if used for other things. This is known in the literature as 'asset specificity'. There is thus a risk involved in the purchase of specific assets, in the presence of uncertainty about the future, that they will not be able to be used as planned and will have a lower value in any alternative use. However, transaction cost analysis suggests that if project-specific plant and equipment become relatively expensive in relation to a project or a firm's overall costs, firms should either integrate the function of plant ownership (exerting maximum control) or redesign the tasks so that general purpose assets will suffice. The UK construction sector has generally adopted the second approach due to the intermittent use of equipment. Consequently, main contractors own only general assets.

Coombs, Saviotti and Walsh (1987) point to several weaknesses in Williamson's approach to the 'in-house' strategy for reducing transaction costs. For instance, they argue that Williamson ignores the conflicts that exist within organisations between departments and between separate subsidiaries. Williamson has been accused of seeming to suggest that, once activities have been brought in-house, it is relatively easy and costless for the central management of the firm to resolve disputes and conflicts of interest between departments of the firm 'by fiat' – i.e. by simple exercise of managerial authority. Coombs *et al.* in effect suggest that this may not be so. This is especially relevant in the construction sector where many of the largest firms own subsidiaries trading with one another as well as with the outside world, such as plant hire firms and contractors. These contractors try to insist that their own plant hire firms must compete on price with outside firms in order to win in-house contracts. This ensures that the main contractor obtains the lowest price for hiring plant. The plant hire department, on the other hand, may try to insist that the contractor always uses their plant when available, so as to increase the plant utilisation rate, the key variable affecting profitability of plant ownership. Each department may conceal information from the central authorities of the firm, and may try to use *political* influence within the corridors of (corporate) power to get a ruling favourable to their interests even if it is not strictly rational from the point of view of the interests of the firm as a whole.

In effect, the central authorities within a large firm face the same two key problems that face central planners in a communist economy – problems of information and motivation (Bowles and Edwards, 1993). Information possessed at lower levels in the hierarchy of a firm will only flow upwards in an imperfect and selective way. Each department or lower level in the management of the firm will try to prevent information unfavourable to its interests from reaching the centre. Alternatively they can try to deluge the centre with more information than it can handle, thus hiding their needles in the resulting haystack.

In order to motivate managers at the lower levels of an organisation, the central authorities often set them performance targets for their part of the firm, usually in terms of profit targets, but perhaps also turnover or cash flow targets. The former requires that quasi-markets (also called internal markets) are established to value transactions between departments of the same firm, so that profit achieved can be measured separately for each. It also means that lower level managers have an interest in manipulating the figures measuring their performance, by understating costs or exaggerating revenues, or by trying to get some of their costs re-allocated, in the management accounting system, to someone else.

Thus much effort is *wasted,* from the point of view of the firm as a whole, in each department trying to improve the appearance of its performance at the expense of other departments of the same firm. All the problems of 'self-seeking with guile', and of information impactedness may reappear in the *transactions* within a single firm. One of the major tasks facing central management of a firm is to try to instil a *collective* ethos amongst its management staff, whilst at the same time retaining measures of individual and departmental performance.

Subcontracting in construction – causes and consequences

We can now combine our discussions of transaction costs and economies of scale and apply them to the construction industry. It is possible to see that large firms in the construction industry take advantage of economies of scale and reduce their transaction costs, and that advantages and gains are not shared equally by both parties to an agreement. Apart from small scale domestic maintenance, repair and improvement work, economies of scale in an industry like construction tend to favour large firms. Small firms, however, survive in the construction industry because large firms find it to be to their advantage.

Thus, in specific construction markets small firms continue to operate. Markets for jobbing plumbers and very small works such as minor domestic housing repairs continue to be served by small firms negotiating directly with their customers. However, on larger projects it is possible to distinguish the advantages of economies of scale and of integration within the hierarchy of a single firm, from the benefits of subcontract markets, in which a main contractor uses separate smaller firms to carry out the work.

Economics of joint ventures

Consider projects that are significantly larger than the average project in the average firm's portfolio, though with project values less than the value added or turnover of at least some of the largest construction firms. We need to distinguish carefully between economies of scale of production, economies of scope of production, pecuniary economies of scale, and risk management economies.

The *scale and integration* of production at the level of the single project can be measured by the value added by a firm, f, on project, p, and this can be less than or equal to the total value added of the project, p. The *scope* of production at the level of the operating business can be measured by the sum of value added of the firm, f, on n projects. Economies of scope come from some centralised resources deployed over the firm's set of projects. *Pecuniary economies of scale* may arise out of gains from unequal exchange or market power, where bargaining power can be proxied by the gross value of contracts won and in hand. These pecuniary economies relate to turnover and not value added. Finally, *portfolio risk management economies* are measured by the cost of holding financial reserves to cover contingencies. The larger the total portfolio of projects of a firm, the less the cost of additional contingency reserves that it must set aside if it adds one new project of an exceptional size.

Now the existence of economies of scale and intergration favour the firm that is able to risk integrating in-house all production on a large project, *P*, if the size of the firm is sufficiently large relative to the project size, in the sense of portfolio risk management. Project organisation based on an integrated single firm would reap both lower transaction costs and greater economies of scale from project, *P*, than would the combined members of a joint venture or consortium.

Provided there is at least one firm of sufficient size that these advantages are greater for it than the addition to risk from taking an absolutely large single project, such as P, into its portfolio, then this firm should be able to outbid its smaller rivals, whether those rivals bid singly or as a joint venture. This argument demonstrates why, generally, a firm will not share work of a type which it is technically capable of undertaking itself with joint venture partners, and why, on the other hand, when project size becomes very large indeed, this ceases to be the case.

Economics of subcontracts

In order to explain why and when firms choose to let-out part of the work they have won as subcontracts, we need to think in terms of quite different variables. Let us leave aside for the moment those main contractors who are in fact only in a position to supply management services, because of the client's choice of procurement route. That is, we are considering main contractors who face a real choice between doing a certain part of the works involved in their contract in-house or placing that part out on a subcontract.

Now, for the same capital employed in the business, the more a firm follows a general policy of fully subcontracting the work on its contracts, the greater the total value of the portfolio of contracts it can afford to operate. It therefore gains advantages of portfolio risk management economies. Moreover, this larger portfolio may at the same time enable it to reap increased advantages of bargaining power in its dealings with subcontractors on each individual project, giving the firm pecuniary economies of scale.

The question then is: are there offsetting advantages to be had from following the opposite policy and doing more work in-house on each of a smaller set of projects instead? An example, where advantages of economies of scale and integration might be sufficiently large to make project integration the best policy, would be where, for instance, sub-

contractors and main contractors on a project had the technical possibility of sharing certain common facilities (site huts, scaffolding,
storage, equipment), but for transaction cost reasons found it expensive to arrange this. Subcontracting in this case would then require
duplication of these facilities and would be a technically inefficient
solution in terms of production costs.

Another interesting case to consider is the design and build contractor. If a firm has won or can win a set of design and build contracts,
the choice it faces is whether to integrate in-house both design and
construction activities or whether to subcontract one or both of them.
It must bear in mind that the more it subcontracts, the greater can be
its ratio of turnover to capital employed. It may be that efficient scale
and scope of production is less in architecture than in construction
management, for example. But it might also be that increased bargaining power *vis-à-vis* manufacturers plays a crucial role and that this
pecuniary economy of scale is proportional to turnover rather than to
in-house value added. Or it might be that the design and build contractor can reap bargaining power benefits from dealing with an architectural subcontractor.

Concluding remarks

This chapter began by examining different methods of co-ordinating
production. Production may be co-ordinated through the management
of the process by individuals who have authority over others by virtue
of their position within the firm and the contract of employment of
their employees. Alternatively, production may be co-ordinated using
markets to select outside firms to carry out the work under the terms
and conditions laid down in a commercial contract between a main
contractor and a subcontractor. However, when market transactions
take place between firms additional costs are incurred by the participants due to uncertainty, lack of information, and the fear that each
may behave opportunistically.

In construction, these transaction costs extend to the costs of the
tendering process itself, and include the design of contracts and the
costs of selecting and later monitoring firms' performance. Moreover,
additional costs arise in the expensive process of specifying products
exactly.

In view of the existence of transaction costs and economies of scale,
the question is: why do firms not integrate the process? One answer
lies in the fact that the main contractors do not *want* to integrate

Table 6.1 **Matrix of market types by number of buyers and sellers**

	Many buyers	**Few buyers**
Many sellers	Symmetrically competitive market: examples include markets in which securities or standard commodities are traded	Monopsonistic market: examples include markets in which large firms buy from subcontractors
Few sellers	Monopolistic market: examples include markets in which major firms sell branded consumer goods and services to a mass of individual households	Countervailing-power market: examples include much public sector purchasing, sales of branded consumer goods by their manufacturers to major retailers, and much corporate purchasing of specialised capital goods

In order to produce the matrix of Table 6.1, we needed the following distinctions:

(a) between standard and branded goods and services;
(b) between consumer and capital goods.

Consumer goods are, by definition, sold ultimately to a multiplicity of small, individually insignificant buyers. If the goods are *branded* goods, a small number of individually powerful sellers have managed to differentiate their products, in the eyes of consumers, from those of their rivals. Branding is the term for the more general phenomenon of product differentiation as it affects consumer goods. Standard goods are simply those for which no such differentiation between sellers' products exists. Capital goods, comprising machinery, plant, vehicles and buildings and works used as means of production, are, again by definition, bought by forms, ranging from very small enterprises to large corporations. Most often, in fact, capital goods are bought by large firms, and each buyer's order can be large relative to the total annual output of each seller. Often, though not always, capital goods are bespoke-made, that is, made (or at least adapted from a standard) to the specifications of the buyer. Though makers of capital goods will try to differentiate their products, they cannot use the mass advertising techniques applicable to consumer goods.

Neo-classical theory of markets

Information is essential for understanding market conditions and making decisions. Some information is available on rival firms. The value and volume of sales for a product type as well as the value and volume of output of an industry may also be published at regular intervals. However, the central piece of economic information sent out by markets is price. Changes in prices signal changes in markets. Prices are formed by market forces. Market-clearing price is the price at which supply is equal to demand. Price changes if either demand conditions or supply conditions change, generating a new demand or supply curve.

The neo-classical theory of markets is therefore concerned with the laws of supply and demand. Supply is defined as the quantity sellers, usually firms, are willing and able to offer for sale in a given period. Effective demand is the quantity buyers, sometimes consumers but sometimes firms, are willing and able to purchase in a given period. The assumption is that both demand and supply quantities hypothetically offered or asked will be functions of price, but with opposite signs. Thus, as price goes up the quantity supplied will rise, while the quantity demanded will fall. There will then be one price at which the two quantities are equalised, and this is the price that the market will find and at which exchanges will take place. All the other price and quantity combinations on the supply and demand curves are purely hypothetical statements of the quantities that would have been offered or asked if price had been other than it actually was. The demand-and-supply diagram involves no time dimension. Changes occurring through time, such as changes in incomes or methods of production, that will alter the quantity demanded or supplied at any one price, can only be represented by a shift to a new diagram. Prices, on the other hand, are imagined to be set instantaneously, without passage of time, and to adjust instantly to any change in demand or supply curves.

In neo-classical economic theory the laws of supply and demand are timeless and universal, regardless of the political, legal, social and cultural context or the level of economic development. Neo-classical economic theory has succeeded in creating a logical mathematical framework to show how prices are arrived at in a perfect market. This theoretical framework is called the price mechanism.

Markets are places where buyers and sellers meet. The forces of demand are manifested through the behaviour of the buyers, while the forces of supply are expressed by factors influencing sellers. The market is where their transactions take place. Buyers seek the lowest price, while sellers seek the highest. In Figure 6.1, assuming all else remains the same, the demand curve shows that the lower the price, the more will be demanded: the higher the price, the more will be supplied. Expressed in the simplest possible form these are the so called laws of demand and supply. The diagram should be read from the y-axis onto the x-axis. That is, it is used to answer the question: if price were to be y, what quantities of x, would be demanded and supplied? The quantity demanded at price y would be x_1 units and supply would be x_2 units. The answer can be easily seen. It makes no sense to try to read it in the other direction, i.e. 'if quantities supplied and demanded are x_1 or x_2 then what will be the price?'

It does not always follow that more goods will be exchanged, the lower the price. Neo-classical theory recognises that supply and demand curves shift in order to take account of changes other than price. If at any price, more or less would be demanded than before, then demand must have shifted. An increase in demand at the prevailing price shifts demand (the whole demand curve or schedule) to the right; a decrease shifts demand to the left. A demand curve would shift if there were a change in unemployment, a change in consumer confidence, or a change in expectations concerning future price rises or falls. Similarly a supply curve would shift if there were a change in

Figure 6.1   Supply and demand curves

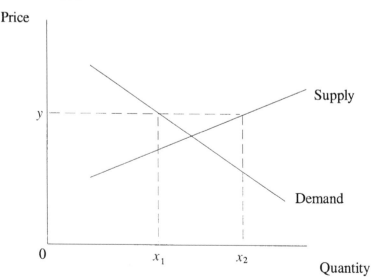

costs, the arrival of new technology, new products or new firms entering a market. For these reasons it is possible for more goods to be
bought (and sold) whilst price increases and for fewer transactions to
take place as price falls.

Market price varies until supply equals demand, when the forces of
supply and demand are said to be balanced or in equilibrium. Hence
the equilibrium price occurs when there is no pressure or tendency for
the price to change. This occurs when there is neither a shortage nor a
surplus. A shortage occurs when more is demanded than supplied at a
given price, whereas a surplus arises when unsold stock cannot find
buyers at prevailing prices. If we allow a small dose of realism, so that
we start from non-equilibrium and recognise that a short period of
time may be needed for the market to adjust towards equilibrium, then
rising prices indicate shortage in a market. Similarly, falling prices are
evidence of surpluses in a market. According to neoclassical theory,
prices do not tend to rise or fall indefinitely but only until a new equilibrium price has been reached.

However, the forces or conditions of supply and demand are themselves constantly changing as a result of exogenous factors. These
exogenous factors include changes in the size, composition and distribution of population, changes in income and employment, changes in
tastes and technology and changes in politics and international trading
conditions. Even changes in the weather, climatic conditions and the
seasons affect market conditions. As a result of these influences on
supply and demand, no sooner has one equilibrium price been reached
than the forces shift, causing further and continuing changes in price.

Alfred Marshall likened supply and demand to the blades of a pair of
scissors. Just as the cutting action of scissors depends on both the
blades, so both supply and demand are equally essential to the setting
of price. In the short term, however, demand is equivalent to the
cutting edge, as it can alter rapidly in response to changing conditions.
In the long run, though, supply becomes the cutting edge as supply
determines what is available on a given market and cost of production
determines price, although changing production methods and output
takes time.

There are several difficulties with this model of markets. Firstly the
demand and supply curves, as we have seen, represent hypothetical
alternative combinations of price and quantities at the same moment.
Only one price and quantity can actually be demanded and supplied at
any one time. All other points on the curves are hypothetical. The
shape of demand and supply curves, therefore, cannot be known by

observation, but only by inference. It is exceedingly difficult to refute empirically any statement about the shape of a supply or demand curve.

Also, the terms demand and supply are left rather ambiguous with respect to time of decision and uncertainty. Is supply, for example, the quantity that sellers 'bring to the market' before they know what the market price that day is going to be, some of which they may then decide not to sell when they discover the market price; or is it the quantity they actually sell; or, the quantity they would have offered for sale had they known in advance what the market price was going to be? This problem arises once we assume that production must precede supply, and that what is offered for sale in the market at any one time has already been produced. How have suppliers decided how much to produce, in advance of knowledge of the market price? Likewise with demand, many orders to buy by firms are in fact made by buyers already contractually committed to fulfilling certain production plans. In order to do that, they must have certain quantities of inputs (and it is now too late for them to substitute another input). These production plans may have been made using expectations about what the relative prices of alternative inputs might be, but must now be fulfilled irrespective of actual prices when the purchases have to be made. Is demand, then, the quantity that buyers have committed themselves to purchasing based on their expectations of the price, or the quantity that they would have chosen to purchase had they correctly predicted the price? Moreover, if sellers or buyers expect price to be high or low, then this will influence the quantities which traders start by trying to place on the market as well as the initial volume of demand and hence their expectations will influence the actual price.

At best supply and demand analysis describes a price and quantity transacted in a given set of conditions at a particular moment. It is a static analysis. Yet the diagram is used to show how surpluses and shortages are eliminated as price moves towards the equilibrium price. The elimination of shortages and surpluses is a dynamic process taking place over a period, while the rate of change is unknown. Actual adjustments to supply, for instance, involve firms changing their production volumes. In other words the analysis is analogous to taking two still photographs of a runner at different points in time and attempting to describe the process of moving from one to the other, by inference. The mechanism is not explained by saying what might have occurred. The process cannot be isolated from the context in which it is taking place without guesswork.

Nevertheless many economics textbooks adopt the neo-classical approach to the price mechanism. For a description of the neo-classical approach applied to the construction industry, see Gruneberg (1997). However, one excellent exception, that is a source of inspiration for much of the discussion here, is Morishima (1984), where he suggests a realistic, time-bound account of market processes.

Flexprice and fixprice markets

To Morishima the way prices are determined varies from market to market depending on the rules for transacting deals, the number of participants in a market and the type of product or service being traded. At one end of the spectrum are fixprice markets, in which price remains relatively constant and quantities exchanged vary. In contrast, flexprice markets are ones in which the price is adjusted until surpluses and shortages are eliminated. In fixprice markets, sellers set and announce their selling prices. In flexprice markets, there is an actual or virtual exchange in which orders to buy and to sell are cleared. Prices are announced by brokers or auctioneers. In fixprice markets, sellers periodically reconsider their selling prices in the light of information about their costs and the volume sold. In flexprice markets, brokers continually readjust their estimate of the market-clearing price, so as to keep quantities supplied and demand in continual balance. Any final product must pass through several interacting layers of market types, some fixprice and some flexprice, as it progresses from its raw material stages to its finished state.

Figure 6.2   Range of market types

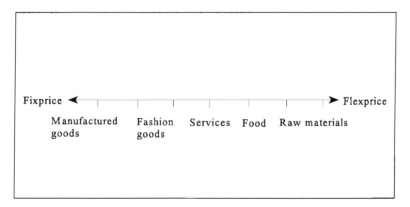

Markets or exchanges for many raw materials are flexprice. Prices respond to the interaction of many buyers and sellers. Indeed commodity prices are in a constant state of flux. However, Morishima (1984) points out that for most manufactured products no exchanges exist where prices are strictly determined by supply and demand.

Manufacturers do not alter their prices to the same extent as commodity prices change. Okun (1981) points out that when there is excess capacity in a market, prices may not necessarily come down. Markets for mass produced durable manufactured goods such as cars, electrical products, furniture and building materials are usually fixprice, prices being set by the supplier during the production process before the goods are sold. Similarly, in markets for services such as railways, hotels, cinemas or hairdressing, prices are set by suppliers before work commences. These markets can also be called fixprice, though that does not mean that prices do not change. It simply means that firms set their prices and it is up to buyers to respond. If firms fail to sell in sufficient quantities, they reduce production in preference to lowering price. If on the other hand, sales increase, rather than change an advertised price, such firms usually prefer to increase production plans. In manufacturing industry it is possible for firms to adjust production to meet demand. The range of flexprice and fixprice markets is summarised in Figure 6.2.

Flexprice *economies* are characterised as ones in which all or the vast majority of markets behave as in neo-classical economic theory. Prices continually vary to reduce shortages and surpluses. Buyers and sellers continually respond to changes in price through competitive trading. This form of market is best illustrated by the stock and foreign currency exchanges, where prices are constantly being adjusted to equate the forces of supply and demand. In stock exchanges, prices are constantly changing and flexible; hence the term, flexprice. Flexprice economies are where the prices of all commodities are decided by competitive trading as in Walrasian economics.

Fixprice economies are where the prices are determined by the full cost principle. Once costs are determined, prices are fixed independently of short-run excess demand for, or supply of, the products.

We can see therefore that prices are not all determined in the same way. In contrast to the neo-classical model, which assumes that prices are continually changing to reflect changes in demand and supply, manufacturers set the prices they charge and then vary their production levels to meet demand. On occasion they work overtime; at other periods they have spare unused capacity. Stocks can be increased or run down to meet short run changes in demand. In industrial product markets supply is equated with demand through changes in quantities rather than changes in price. Price is not used to clear the market, unless unplanned stocks of unsold goods remain and cannot be stored. Competition between firms can on occasion lead to price wars but, as Morishima (1984) points out, these mostly occur as firms strategically attempt to weaken competitors rather than through oversupply of goods.

In fact the characteristics of flexprice and fixprice markets are combined in real economies, where markets for commodities tend to be flexprice and markets for manufactured goods tend to be fixprice. This approach to market theory still allows markets to vary in terms of the number of participants, their location, the area they cover, the degree of competition, concentration of ownership and the rules for the conduct of business. Markets are places where the participants interact to resolve their different economic interests. Sellers still seek to obtain the highest prices they can for their goods, while buyers still want to purchase at the lowest price.

There is no reason why all prices at all times in all places and in all commodities, products and services should be arrived at in the same way. It is not that neoclassical theory of price determination is completely wrong; it remains useful and essentially correct as an abstract model but for only one set of market conditions.

In construction it is not a simple matter for a firm of announcing a 'price' and then of reducing or expanding construction work, because the work load of a contractor is determined by both the invitations to tender they receive from clients and their choice of tender prices for each project. Moreover, because of the size and proportion of work represented by each project any adjustment to work load by increasing or decreasing the number of projects can be more than 'marginal', with significant implications for a firm's cash flow and plans.

Of course, this simple dichotomy between fixprice and flexprice does not cover all cases. Where one-off construction components or finished projects are concerned, *negotiation* between a single supplier and a single buyer will often determine the price on the basis of the personal negotiating ability and bargaining strengths of the participants. This represents a third form of price determination.

While an ability to negotiate depends on the personality of the participants in the process, their bargaining strength depends on a variety of factors including the durability of the supply, its degree of technological sophistication, the level of demand for the suppliers output, the substitutability of the product, the urgency of the supply and the dependence of a larger scheme on the successful outcome to the price negotiation and supply of the product. Bargaining strength also depends on possession of information. For example, if a buyer knows the minimum acceptable prices of the two lowest-cost potential sellers, the buyer can negotiate a price no higher than the minimum acceptable to the second-lowest-cost firm (see Vickery, 1900).

A market in construction contracts

In markets with small numbers of buyers and sellers, the sequence of transactions can affect prices. This is important in the case of construction. To illustrate the point, and its application, imagine a market for units of a homogeneous constructurion product, comprising a given number of potential buyers and a given number of potential sellers. Each buyer announces their 'offer to buy' in the form of an organised tender competition (a 'reverse auction', with one buyer and competing sellers) with the order awarded to the lowest bidder. Further, imagine that the price and identity of transactors of each agreed transaction are public knowledge, and that offers to buy are dealt with in a randomly-ordered sequence. Each seller keeps private their own supply schedule of the quantities they would wish to sell at each price. Each buyer is publicly known to be in the market for a certain quantity. This quantity is announced as part of their invitation to suppliers to bid.

A market clearing price would then be the price at which each seller was able to win contracts just equal to the total quantity they wished to sell at that price. We have assumed that buyers are pre-committed to demanding a certain fixed quantity, in the short run. However, it is perfectly likely not only that many transactions will be made at prices above or below the market-clearing price, but also that as a consequence of this, even the average of actual transaction prices will not clear the market, in the sense that, at the end of trading, there will be some sellers who would have been willing to sell and produce more than they have actually contracted for, at that average transaction price.

This result arises from three institutional features of the market for construction contracts:

- the fact that they are let in sequence,
- the fact that contracts cannot be retracted or assigned, that is, sold to another party, and

- the number of bidders for a project is arbitrarily limited to a relatively small number of firms, which represent only a fraction of the total number of potential suppliers.

In the absence of any better guide available to them, let us assume firms begin their tendering in a particular period by taking the level of bid prices that they used in the previous period, and adjust this upwards or downwards according to whether, in that previous period, they felt short of or exceeded their desired volume of sales. Let us also assume they then make a further adjustment based on their view of how demand and the strength of competition from rivals have changed between the two periods.

Using this method of arriving at a level of prices, they commence bidding for the first projects offered on the market by buyers. After, say, one quarter of all projects have been bid for, each firm reviews its progress. Some find that they have won more projects or made a greater value of sales than they expected. They may find they are on course to exceed their desired level of sales for the period as a whole. These firms most likely respond to their situation by raising their prices for the next tranche of projects offered on the market. Other firms find that they are on course to fall short, and probably reduce their prices for the next set of bids.

Meanwhile, let us look at things from the point of view of a client, a buyer for one of the first projects in the sequence. The buyer selects at random a small number of suppliers to bid. Each bid is then *sealed* in that it is made known to the buyer or the auctioneer but not to the other potential bidders. Bids must be submitted by a closing date, when bidding is declared over, and all bids received are compared. The project is then awarded to the lowest bidder.

Now, first, it is clear that it is possible that amongst the non-bidders there may be several who would in fact have been willing to undertake the project at the lowest bid price, or even somewhat below that, and who know this immediately the winning price is announced. Second, some of the actual bidders may have submitted prices higher than the minimum price at which they would have been prepared to accept the contract. They gambled that they could still win the project at a higher price and preferred a lower probability of making a large profit from the project, to a higher probability of making a small profit. Early in the bidding season this behaviour is more likely, because if they lose their bet, there is still plenty of time to adjust and make up their target of sales on later projects. Third, we can assume that all bidders are aware that all estimates of prime cost are subject to error, and that there is therefore a potential problem of *winner's curse*. The winner's curse is that the lowest bid is likely to be from a bidder who has under-estimated cost; *ex post* profits on projects are therefore systematically less than expected *ex ante* mark-ups (Fine, 1975). However, to avoid winner's curse, some firms will have added an error-protection mark-up to estimates, but the additional mark up will have lost them the work. When the lowest-bid price is announced, some such bidders may come to suspect that such has been the case for them. They will only know this, however, if and when they discover the winning bidder's out-turn costs.

We have, therefore, in fact, four sets of bidders all expressing *regret* that they did not submit lower bids. Two of these feel regret immediately, the third when the *true* costs of the project are revealed. The fourth, not yet considered, consists of those firms who, at the end of the period, find they would wish to turn back the clock and revise their general level of bid prices downward. At the end of the bidding season, the client too may have cause to experience regret. It may transpire that subsequently-let similar projects were placed at lower prices. Finally, the winning bidder may express regret that they did not submit a higher price, either because their bid was lower then it needed to be to win the contract, or because they find themselves victims of the winner's curse.

Non-assignment of contracts is important because it prevents suppliers from *hedging* their market position, and it introduces an asymmetry into firms' responses to the evolution of prices in the market. If assignment were allowed, then a firm could sell (to other contractors) the right to perform contracts won at higher process than are now current (for an amount equal to the difference in prices). Moreover, buyers offering new contracts on the market would always be in competition from construction firms who had won earlier contracts which they would be prepared, at certain prices, to assign. In effect, assignment would permit professional *dealers* in contracts to emerge, and the activities of these dealers would tend to both stabilise prices through time and cause closer clustering of contract prices one with another.

Dealers would *buy* contracts at any price below their estimate of the price at which they could resell them – and the latter would reflect incorporation of learning and response to regret by individual clients and contractors. Just as stock markets have settlement periods, the dealer would estimate the end of period, market clearing price. Unless the number of sellers was very small, this market clearing price would depend very little on the market *position* of any one contractor, unlike the prices offered directly by contractors themselves. It is the absence of dealers more than any other single thing that causes markets for construction contracts not to clear and causes trading to occur at divergent prices affected by chance factors and by place in the sequence.

The conventional spectrum of neo-classical market types

Of course, economists have long recognised that not all markets are identical. Figure 6.3 illustrates the conventional view that markets form a continuous spectrum of varying degrees of competition ranging from perfect competition to pure monopoly and monopsony.

At one end of the spectrum in Figure 6.3 is perfect competition in which there are many buyers and many sellers. In perfectly competitive markets there are no barriers to entry or exit. Anyone who wishes to trade may do so. However, for the buyers and sellers to be engaged in perfect competition with each other, they would all need perfect knowledge. They would therefore have access to all relevant information regarding prices and costs, technology and availability of materials. Only on this basis would competition be able to ensure efficient production by the suppliers and the maximum satisfaction of all buyers. Only then could the optimum distribution of resources be guaranteed by the market.

If a market fails to satisfy absolutely all the conditions of perfect competition, then that market is said to be imperfectly competitive. A

Figure 6.3   Different types of competitive and non-competitive markets

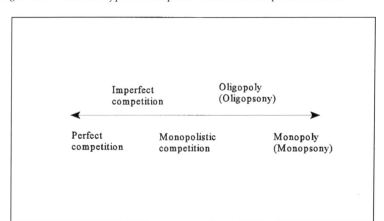

market with many buyers and many sellers and no barriers to entry or exit, for instance, but without perfect knowledge on the part of all buyers and sellers would therefore be imperfectly competitive.

However, our approach is to stress that, in fact, the predominant market form in developed economies is oligopolistic competition, in which sellers set their prices, and in which, consequently, prices are rather fixed through time; but that alongside this there exist markets of an opposite kind, by no means necessarily 'perfect', in which each seller has to take the market price, and in which prices fluctuate sensitively to demand and supply conditions.

The concept of perfect competition has had the consequence of making economists think of *degrees* of imperfect competition, which naturally leads to the idea of a continuum. The virtue of the fix-price/flexible-price distinction is that it focuses on a significant duality. Its other great advantage is that it fully reflects the two-sidedness of markets (sellers and buyers), whereas the language of perfect competition, imperfect competition, monopolistic competition and oligopoly tends to stress the supply-side structure – the number of sellers.

An oligopoly is a market dominated by a few sellers. Each firm attempts to differentiate its products and services from those of its rivals in order to ensure its share of the market. At the same time, each firm is aware that any actions it takes will have an effect on its competitors who will in turn respond in kind. In oligopsony, only a few firms make purchases. A local labour market with only a few firms taking on labour would be an example of an oligopsonistic market.

In perfect competition the number of firms or sellers is so great that anything an individual firm does will only represent a negligible proportion of the market and have no effect on rival firms or price. In contrast, oligopolistic competition involves prior consideration of the response of rivals before changing prices, output or introducing new products. Each firm represents a threat to its rivals as they battle with each other for market share and domination. This has given rise to an interesting field of application for 'game theory'.

Theory of the kinked demand curve

Perhaps the most simple yet persuasive answer to why fix-price markets exist is that first offered by Sweezy (1939), now known as the theory of the kinked demand curve, AZB in Figure 6.4. Because of the relatively few major competitors, firms in oligopolistic competition are reluctant to raise their prices above the level of their competitors. Hence the elastic demand above the kink, AZ on demand curve D_1 in Figure 6.4. The effect would be a reduction in market share as the other oligopolists would leave their prices unchanged. Rivals' products would then become relatively cheaper and take customers away. A reduction in price on the other hand would only result, the firm predicts, in similar reductions by competitors, ZB on demand curve D_2 in Figure 6.4. There may be a slight increase in sales as a result of lower prices but the firm's share of the market would not tend to rise. Lowering prices would at best simply reduce revenues; at worst it would lead to a price war with further rounds of damaging price cuts.

For these reasons firms in oligopoly are reluctant to compete on price. That is not to say that oligopolists do not compete. Non-price competition between oligopolists can be fierce and costly. After sales service, advertising, sales promotions, sponsorship, packaging and public image are all used by firms to capture market share, to say nothing of the continuous search for product innovations.

In fact the neo-classical 'rule' for profit-maximing, (Select output at which $MC = MR$) cannot show how oligoplists make decisions: around the kink of the demand curve price and output is held constant even when marginal costs and revenues rise or fall between *a* and *b* in Figure 6.4. The diagram has nothing to add to the strategies open to oligopolists. It only shows the constraints on firms, which cannot or dare not change price or output in any given situation. It does not explain how firms actually behave and the options open to them.

Figure 6.4   The Sweezy diagram – oligopoly

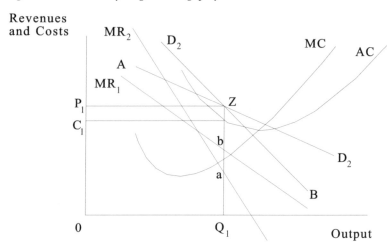

When a set of oligopolistic rivals' costs all go up, for example because of an increase in input prices, oligopolists will have no hesitation in passing these on as higher prices, because they know rivals will have experienced the same cost increases. In this case, when the market leader increases its prices, the other firms are almost certain to follow. However, if most inputs are purchased in fixprice markets (e.g. if labour wages are set by annual national collective bargaining, whilst fuel and materials are purchased on term-contracts), these cost increases will only occur periodically, and in between times prices stay fixed. What firms are actually thought to try to hold constant in fixprice markets is their profit mark-up. The idea is that there is a customary average mark-up which each market has established over time, and there is a tacit agreement not to deviate from this mark-up unless there is a major change or *economic shock* which permanently alters the balance of the market. A firm with higher than average costs may add a lower than average mark-up, to keep its prices competitive – but each firm knows what its *normal* mark-up is, and is only likely to depart from it if it thinks there has been a fundamental and lasting change in the underlying balance of demand and supply conditions. In response to anything less than that, the firm lets its volume of output and sales adjust to absorb changes in demand. All this implies 'well-behaved' markets, where conditions do not change very quickly. It also implies that firms normally possess unused spare capacity which they are anxious to utilise whenever an increase of demand permits.

Theory of contestable markets

One recent modification to the traditional neoclassical approach to markets, and another basis for explaining the existence of a fixprice sector, uses the concept of contestability. A contestable market, according to Baumol (1982), is one in which the competing firms moderate their behaviour because of the *threat* of attracting potential competitors into the market. For instance, if it became widely known that profits in a particular location were high, then new firms might begin to advertise their services in the area, under-cutting existing prices to win customers. The contestability of a market depends on the ease with which new entrants can obtain a market share. If *exit* from a market is costless, due to low asset specificity, then even ephemeral super-profits are unlikely. They are prevented by the threat of hit and run temporary entry.

Where markets are highly contestable, existing firms will tend to charge prices and provide a quality of service similar to what the potential competition would charge. Indeed the threat of competition might even induce firms to lower their prices to make it unattractive for new entrants. In other words it is the threat of new entrants, rather than actual new entrants which determines the pricing and offered output behaviour of firms.

Although this approach is still firmly rooted in neo-classical economics, it is quite consistent with the findings of some empirical research. Neale and Haslam (1994) report on the work of Andrews and Brunner, who found that firms often fixed their prices just low enough to deter new entrants into the industry. They only raised their prices when they had to in order to maintain their profit margins in response to an increase in their costs.

If there are significant barriers to entry, then firms need not fear increased competition from outside. Prices would be higher and quantities more restrictive than in a more contestable market.

The contract construction market

Although there are many thousands of firms in the contract construction industry, the industry is fragmented regionally, locally and according to specialist markets of subcontractors. In any one market, leaving aside the markets in which the smallest firms operate, only a relatively few firms dominate and some control a significant proportion of market sales. Moreover each firm is well aware of who its close rivals are, and monitors their behaviour and success carefully. Much the same is true of markets for professional services in construction. In both cases, geographical barriers to entry are sufficiently high to exclude non-local firms from the local market for all but the largest projects.

However, contract construction also provides good evidence for the point that oligopoly is not by itself necessarily sufficient to prevent the occurrence of price competition and flexible prices. In construction demand booms, contractors' prices are far from stable. Construction firms tend to raise their prices even before their costs begin to increase, in order to take advantage of a *sellers' market* and to widen their profit mark-ups. Similarly, however, in demand slumps, contractors' prices can fall more rapidly than the index of construction costs (made up of building materials' producers' list prices and wage rates for directly employed labour). Profit mark-ups can fall towards zero per cent and in some competitions for work firms have been known to put negative mark-ups (on costs estimated at current levels of input prices) in order to 'buy' work to gain some cash flow and turnover. In fact, when a negative mark-up has been used at the tender stage, construction contractors, in order to regain profit margins, put added pressure on subcontractors and labour to reduce their charges and wages.

Why do contractors not keep their mark-ups constant, in response to what are usually only short-lived demand changes, in the way that fix-price theory predicts? The answer lies partly in the fact that they do not buy their inputs in fixprice markets (e.g. labour only subcontractors' wages fluctuate massively and quickly in response to market conditions, unlike wages under national collective bargaining), partly in the fact that contractors do not really carry expensive spare capacity, and partly in the way that (often oligopsonistic) construction clients take advantage, in a way that ordinary consumers could not, of recessionary conditions by deliberately stimulating tender price competition whilst contractors have to wait for boom conditions to restore margins.

From the market for a single construction project to the market for a type of project

Assume that each single project constitutes a separate market. We can then derive a clear and meaningful definition of *market clearing* price. It is simply the price at which the quantity of *willing* supply from one firm equals the quantity of demand. It is the price at which all bar one potential supplier choose to withdraw from the market. However, only if certain special conditions hold will a simultaneous sealed bid tender competition generate this same price. We take the view that, by and large, these special conditions do not hold, and hence, our notion of *ex post* bidder or client regret.

Even more fundamentally, moreover, we question the idea that it is satisfactory to think of each single project as constituting a distinct market. Standard industrial economics teaches that the boundary of a market is best given by the discontinuity in elasticities of substitution. That is, within a market the cross elasticities of demand and supply are significantly higher than they are between that market and other markets. Put concretely, a set of construction projects constitute a market if a change in the price of one or some of these projects will cause a significant change in either the demand for or supply to the other projects in the set: whereas it will have no significant or less measurable effect on demand or supply for projects outside this set, in other markets.

Let us consider new trunk road projects as an example of a possible construction contract market. Suppose that prices for one set of projects within this market, all placed recently, have gone up and have been higher than anticipated. This may occur, say, on contracts for the sections of a major new motorway. What effect will this have on (a) demand for and prices of other trunk road projects, and (b) demand for and prices of construction projects of other kinds?

Considering first the *demand* side, it might well be that the client, in this case the Department of the Environment, Transport and the Regions (DETR), responds to these higher prices by postponing, or scaling-down its specification for, new road projects being prepared to be put out to tender. This would be an indication of a cross-elasticity of demand.

The standard notion is that prices are known before buyers make their decisions. If the price of X rises relative to that of Y (a close substitute in the eyes of customers), then buyers react to the rise in the price of X by buying less of X and switching some of their demand to Y. In our example, in contrast, the increase in the *ex post* price of X has no effect on the demand for X, because the DETR are pre-committed to place projects to the lowest tenderer, however high that tender turns out to be. Instead the increase in the *ex post* price of X reduces demand for closely similar projects Y! This is because we are considering the *income* or *budget* effect **cross-elasticity** of demand, not the relative price effect. The increase in expenditure on X reduces the balance of income or budget available to be spent on Y. This effect depends on X and Y having the same buyer and being purchased from a common budget. The other reason for the effect of the price of X on demand for Y is that the client uses the observed price for x as information on the expected price of y.

Relative-price effects, if any, are likely to show up over a rather longer time period. In this case, a rise in trunk road construction prices might conceivably cause the DETR to switch some of its capital budget away from future new roads projects and towards rail projects or improvements to existing roads. However, we suspect that, both in this example and more generally, **relative-price** cross elasticities of demand between construction projects are not normally very strong, and are often outweighed by **budget** cross-elasticities.

Meanwhile on the supply side, the question is whether these higher prices for some road projects will draw-in new suppliers to other, later road project bidding? What are the barriers of entry to the road projects market? The advantages to incumbents with investments in dedicated plant and equipment or in skills and experience are, in this case, almost certainly quite large. Moreover, the whole system of pre-qualification of would-be tenderers tends to exclude firms which lack recent prior experience of doing similar projects.

On the other hand, let us suppose that prices for road projects in one region of the country were found to be rising in relative terms, because of a clustering of demand in that region. One would then expect to find roadbuilding firms switching their tendering efforts (and, subsequently their production capacity) from other regions to this one. In other words, the market for trunk road projects is a national one, because the cross elasticity of supply between regions is highly responsive to regional price differentials.

The single project is not a market in itself. For this reason construction contract prices can not be modelled adequately as being set in flexprice markets. Instead we should imagine markets consisting of limited numbers of suppliers engaged in fairly direct rivalry. This rivalry can be measured, for instance, by the frequency with which these firms find themselves competing for the same projects. Each firm operates with a general *ex ante* and perhaps *ex post* rate of mark-up on direct costs. This rate of mark-up is first, characteristic of that market. Secondly, the firm's mark-up rate is dependent upon the relative position of that firm in that market in terms of whether, for example, it possesses relative cost advantages or disadvantages.

Some evidence of the *ex post* similarities of profit margins of firms operating in similar construction markets at various stages in the demand cycle does exist (Akintoye and Skitmore, 1991). However, the evidence is made hard to gather by the fact that profits reported in company accounts normally refer to a firm's operations in several, or even many, markets. For *ex ante* margins one would properly need to know firms' estimates or forecasts of prime costs. We do have several studies of the dispersion or spread of tender prices on some projects (Rawlinson and Raftery, 1997; Skitmore, 1989). These studies reveal a low but far from negligible dispersion of tender prices. Our interpretation of this low dispersion of tender prices is that it can be taken as supportive of the idea of firms all passing on similar cost increases with a narrow range of market-norm mark-ups added.

Mixprice in the market for construction contracts

As we have seen, the construction contracts market has neither the classic profile of a flexprice nor a fixprice market. On the one hand, there is no clear market-clearing price which sellers must take. On the other, mark-ups added by firms to their cost estimates do not have the relative constancy postulated by the fixprice model.

Thus we propose the descriptive term *mixprice* to suggest a market in which we find a combination of (a) the cost-plus-mark-up price setting behaviour found in fixprice markets and (b) the sensitivity of prices to demand fluctuations through time of flexprice markets.

Why do contract prices vary strongly in line with changes in demand? First, the structure of construction firms' costs is dominated more by prime cost than overhead costs. Thus any tendency of prime costs per unit to vary proportionately with demand and output volume might outweigh any tendency for overhead costs per unit to vary inversely with volume. Therefore, average total costs per unit might vary directly with demand volume. A constant net profit mark-up would then yield prices varying directly with demand. However, this line of reasoning may not be very realistic because it implies that firms arrive at tender prices by adding a net profit mark-up to estimated total unit project costs. It is more realistic to think of firms as adding a gross mark-up, including profit plus contribution to overheads, to estimated prime costs. However, it is undoubtedly the case that firms' unit prime costs do change in line with demand.

One way of expressing this tendency of costs to follow demand is to think of price elasticity of supply over time for the whole industry or for a market. The idea of the elasticity of supply over time is that a given change in price causes a certain change in supply in the next time period, then a larger one in the following period, until eventually adjustment to the new price is complete, and supply ceases to change.

Now, in construction the initial elasticity of supply is low, partly because of the lack of efficient mechanisms to transmit the information that price has changed, partly because of the lack of stocks and of ready-to-use spare capacity existing within firms, and partly because of the time required for production-decisions and investment-decisions (in response to the price change) to affect supply. Unitl supply responds, increased demand necessarily bids up prices.

The second class of reason why prices vary in line with demand involves a postulate of variable mark-ups. We discuss this in Part IV at length, in terms of a model of firms with target turnover and real output ranges, where the number of firms is given in the short run. The main forces working to restore normal mark-ups then become the entry or exit of firms, in the longer run.

This is a suitable point to make some general observations about the phenomena of entry to or exit from construction markets. In the markets for very small projects, supplied by jobbing builders, entry and exit of firms is easy, quick and of little cost. Thus it is mainly lack of information spreading to other firms that lie behind any persistence of higher or lower gross margins than in other markets for any firm or set of firms established in a market.

However, in the market for larger projects taken as a whole, neither entry nor exit of firms are very common occurrences. Contractors have no obvious alternative lines of business to follow, thus explaining low exit; whilst potential new entrants seem discouraged by the need to incur sunk costs in order to enter the market. These sunk costs, which cannot be recouped at exit, are essentially the costs of building up expertise and reputation. They are also discouraged to enter by the large number of existing competitors, and unfavourable long term demand trends. Thus it is relatively uncommon for firms to enter construction from outside at a large scale.

What is common, however, is for a construction firm already established in one market or set of markets to try to enter new ones when margins there seem significantly higher, and to withdraw from certain markets when margins there are relatively low. Insofar as this occurs, it will tend to equalise margins between construction contract markets. However, we doubt that it is such a strong phenomenon as to eliminate such differences or to create in effect a single market on the supply side.

New housing market

For the housing market as a whole, the average prices and numbers of houses on the market varies continually. Although house prices vary, they do so in a steadier way than stock market prices taking longer periods to adjust to changing circumstances. Meanwhile, the offered price of a specific house, for instance, is set and then held constant until a decision is taken by the seller to raise or lower it. It is easier for the individual vendor selling one house to adjust the price rather than the quantity.

However, for speculative housebuilders, it is possible for them to adjust output over a period of time rather than alter prices. Alternatively, they may offer completed houses on the market with extras such as fitted kitchens, interest free periods and other inducements to purchase. Indeed this type of rigid pricing behaviour is commonly adopted by managers across the whole spectrum of industrial and commercial sectors, where it is easier to market a 'free extra' than to lower prices offered to new customers, without depressing customers' price expectations, and offending old customers. Price rigidity points to another set of market mechanisms in manufacturing, construction and property.

Building materials and components markets

The building materials industry, unlike most of contracting, is an example of a national oligopoly. For basic building materials like cement, bricks, glass and ready-mix concrete each national economy normally contains just a handful of major producers, traditionally with limited market penetration by imports.

Well publicised examples of explicit price-fixing agreements between oligopolies in construction have included ready-mix concrete (as well as contractors') price rings or cartels. But, though cartels certainly assist firms to prevent price competition, they are not a necessary precondition.

Markets for building materials and components are fixprice markets. The suppliers determine the price by calculating their costs and adding a percentage mark up. The price is not confinally altered in response to changes in demand. Rather, the quantities produced and offered for sale will tend to be adjusted, unless manufacturers are operating at full capacity (see Chapter 7). The index of building materials prices moves closely in line with indices for the general level of manufacturers' direct costs.

Manufactured building materials markets are one of the most important instances of fix-price markets in construction. Building materials prices in fact behave in a way which is typical of a wide range of manufactured goods produced under oligopolistic supply but with many buyers. The few large, powerful buyers of such materials in construction constitute a partial special case. Faced with this countervailing power, the result, equally common in markets with few sellers and few buyers, is a process of confidential price negotiation. The negotiations, in this case, are conducted around the size of bulk purchasing discounts off the published list prices, and the length of credit period before payment.

Property markets

Markets for property, especially the built stock, also have a fix-price character in part. For commercial property such as offices and shops, rents are normally fixed for several years at a time. Moreover, there may even be contractually agreed formulae to determine the amount of upward increase in rents at each periodic review date. This applies to property with existing tenants.

However, rentals on newly built property are much more flexible When demand is slack, there may be a time lag before sellers realise there is little take up in the market but their estate agents working on their behalf will tend to advise on price reductions in order to speed up the letting of property. The vacant stock has to absorb the whole supply and demand change adjustment, so that even large reductions in initial rents may not suffice to balance demand and supply in flow terms.

On the other hand, capitalised market prices for the purchase and sale of freeholds or long leaseholds in this same commercial property are equally clearly part of a flexprice market, and show both large and continuous fluctuation according to market conditions. The property capital market (the market in freeholds and leaseholds) is an example of a flexprice market, in which property prices readily respond to changing demand conditions. At a time of excess demand property prices will rise as competing potential owners outbid each other in an effort to acquire the ownership of a building.

The property market in recession, however, does not reduce its prices to a level sufficient to attract tenants or buyers for all property offered on the market. During recessions many buildings remain empty, in the hope that a tenant can be found within a period of time. Reducing initial rental income reduces the value of property and reductions in rent would therefore lead to reductions in the book value of the assets of a property owning company.

Just how downwardly sticky asking rents will be in a property recession depends to a large extent on the financial structure of the main property developers. If these firms are highly geared in flow terms (debt charges per annum are high as a proportion of total rental incomes from the firm's portfolio of properties), then short-term cash flow considerations may predominate – i.e. some new rent flows must be obtained, on whatever terms, or else the firm faces bankruptcy. In this case asking rents will fall sharply. If, on the other hand, covering debt charges out of rents is not a major problem, (for example, because the portfolio of existing tenanted properties is large relative to the level of recent development actively by the firm), then longer term considerations will most likely predominate. The firm might prefer to lose some rent income now, by keeping properties vacant, rather than lose future rent income and book values by entering into long term tenancy agreements at depressed rent levels. Alternatively, the firm might wish to let them at normal rents but with substantial initial rent free periods. Once tenanted even on such terms, it may then be possible to dispose of such properties to financial institutions, or raise loans on their security.

The interaction between fixprice and flexprice markets in construction

Given that construction contains both types of market, does this enable us to divide it clearly into market sectors? How do fixprice and flexprice construction market sectors intersect within the process of production, and with what effects both during the growth phase of the 30 years of the post-war social structure of accumulation and in the supposed phase of the new post-1970s social structure of accumulation?

These issues can best be approached through a *business cycle* analysis. Economic cycles consist of alternate periods of expansion and contraction of demand and output. These periods are usually referred to as the recovery and recessionary phases of the cycle. The point is that, over the course of a cycle, profits in any part of the economy will in part be determined by the distinctive time-series behaviour of fixed and flexible prices. During recovery phases, flexprices will tend to rise more rapidly than fixprices; but, conversely, during periods of crisis and contraction, flexprices will tend to fall relative to fixprices.

Evidently, any capitalists selling their output in a flexprice market but buying their inputs chiefly in fixprice markets, for example, would find their profit margins widen during cyclical expansions and narrow during contractions.

To understand the position of firms operating in both fixprice and flexprice markets, it is worth noting that the top right and bottom left cells of the matrix in Table 6.2 are not symmetrical opposites. Fixprice, we must remember, does not mean that prices are literally unchanging. Specifically, cost increases will be passed on, as may even cost reduc-

Table 6.2   Profit margins and market sectors

	Outputs flexprice	Outputs fixprice
Inputs flexprice	π constant over cycle	π falls slightly in expansion, rises in contraction
Inputs fixprice	π rises in expansion, falls in contraction	π constant over cycle

tions, though probably to a lesser extent. However, there will be some time lag before this happens, causing a slight squeeze on profits during expansion phases for those firms with flexprice inputs and fixprice outputs. If there were no time lag, the profit margins of those firms would not fall at all.

Now, traditional main contractors and subcontractors are the prime example of construction firms who sell in a flexprice market, but buy a large part of their inputs, materials and components, in a fixprice market. Thus we would expect these firms' profit margins to move strongly pro-cyclically, and this is pretty much what we actually find when we try to test the hypothesis. Management contractors are quite a different case, because they do not buy material inputs, but leave that function to their supply-and-fix work package subcontractors.

Concluding remarks

While neoclassical approaches to competition and oligopoly provide general approaches to markets, they do not explain what actually happens. We have seen that different markets operate in different ways even within the construction and property sector. While the characteristics of some markets lead them to be flexprice markets, others are fixprice, and others mixprice. Tendering for contracts, speculative house building, and construction labour markets all have specific qualities which lead to different ways of doing business, some fixprice, some mixprice, others flexprice. As a result the construction sector is comprised of many interwoven markets. The theoretical implication of this is that as all firms operate in different types of markets, some fixprice and some flexprice, their profits are affected by the type of markets from which they purchase and the type of market in which they sell.

This is a constantly changing relationship over the course of a business cycle. During growth phases, in fixprice markets, prices remain the same to begin with and only rise after a time lag while in flexprice markets prices rise in rapid response to increased demand if supply lags. Consequently firms selling into flexprice markets can increase their profits by raising their prices until fixprice market prices begin to rise and squeeze profit margins. The reverse is the case in recession. During recessions, fixprice market prices remain the same to begin with and only fall, it ar all, after a time lag, in line with costs, while in flexprice markets prices fall in rapid response to declining demand.

Consequently firms selling into flexprice markets find their profit margins decline rapidly, until fixprice market prices begin to fall. The situation is not so extreme for firms which buy in flexprice markets and sell in flexprice markets, or firms which buy in fixprice markets and sell in fixprice markets.

Part III
Capacity

7
Capacity of the Construction Industry

Introduction

In order to plan and co-ordinate the physical activities and output of industries, planners require *volume data,* such as technical input per unit of output. For instance, will there be enough bricks to permit a particular volume of construction? If not, can brick industry capacity be increased, or can construction industry demand for bricks be reduced? Even in a capitalist market economy, the need to ask this kind of question is clear, particularly in a relatively closed sector like construction.

In a completely open economy, if the necessary bricks are not available locally, they can be imported. This shifts the planning issue from the industrial to the macro-financial level. It becomes a question of the impact of imports and the balance of trade on macro-policy variables, like interest rates or exchange rates. However, as we shall see, the construction sector is not yet *open* in this full sense. An excess of construction volume demand over capacity leads to major, harmful effects on building quality, inflation, delays and abortive projects. An excess of capacity over demand, on the other hand, leads to problems of unemployment in specialist occupations and under-utilisation of assets.

In this chapter we consider various attempts to define industrial capacity. We see that the concept is necessarily elusive as it refers to a potential rather than an existing quantity of output. We then attempt to apply a definition of capacity to the construction industry and understand the relationship between individual firms and the capacity of the industry as a whole. We then consider the capacity of the construction industry and its international implications.

The neo-classical approach to capacity

The theoretical position of neo-classical orthodoxy remains what it has always been – that *price* movements, not volumes, are the necessary and sufficient information that firms and others need to monitor, in order to make their own output and investment decisions. Physical shortages, it is assumed, will be self-liquidating, once prices rise to reflect the shortage, thus stimulating increased supply.

There are several problems with this faith in the price mechanism. There is an ambiguity as to what exactly is being signalled by a price increase. Has it been caused by increased demand, or by reduced supply, or by increased costs? Are the conditions behind the price change likely to persist or prove temporary? Have other producers already responded in ways that will result in the price falling again? A further difficulty with the price mechanism is the problem of the short-run, whilst waiting for the time-lags involved in responses to price changes to work their way through. In the interim, the effects of physical shortage or over-capacity can be very serious, either for inflation or for unemployment.

The individual firm in construction

Firms can try to guess what demand levels might result from an infinity of unco-ordinated private decisions. Inability to do this with confidence or success encourages firms to think more in terms of wait-and-see-then-respond and less in terms of anticipation and planning.

Various research bodies publish medium-to-long term forecasts for construction demand volumes, for example, *Construction Forecasting and Research*, as well as some of the more disaggregated macro-modellers, such as *Cambridge Econometrics*. However, the track record of these forecasts has not inspired great confidence in them. Thus, whilst the strategic planners of most big construction sector firms purchase such forecasts and study them quite carefully, few industry decision-makers regard them as a sufficient basis on which to make major investment or disinvestment decisions.

Construction firms produce diverse outputs, not readily reducible to a single physical measure of quantity. However, a firm can be thought of in terms of multiples of a standard 'bundle of resources required' for a level of output. One *unit of workload* is then defined as that which normally requires one standard bundle of resources to execute in a given time period. Expansion of output by the firm can be seen in terms of adding units of workload to the demand placed on resources. Workload needs to be defined in a way that links it clearly to resource and capacity. However, to do this we have had to smuggle in the assumption that the same output 'normally' requires the same resource input, thus disallowing variations in productive efficiency from one project to another, or as the firm gets busier. Thus we have, in effect, had to assume what we would like to test – that unit costs, disregarding price effects, are constant with respect to volume of output of the firm.

In what follows we will speak of 'units' of construction output, 'unit cost' and 'unit price', but always bearing in mind that this is a simplification, which is highly unrealistic. Though it may introduce distortions into our analysis, we do it for want of a better alternative. Towards the end of this chapter the simplifying assumptions will be removed or relaxed.

Output and price determination for natural and produced goods

For *natural goods*, such as land, prices depend in the long run solely on demand. This is because the quantity of supply is already fixed. Because they cannot be produced, there can be no question of a cost of production influencing price. For such goods the supply curve must be understood as a willingness-to-sell schedule, where sellers compare value-in-use or potential future exchange value with present prices in order to determine their willingness-to-sell at each price. Something similar, as we shall see, can be said of any merchant or speculator's supply-response to price level. If natural goods are also durable, as they often are, then movements in the current price depend on the ratio between current and expected future demand, because of the influence of the latter on expected future price. Long run stability of price for such goods occurs when present and expected future demand are in such a relationship that, at the stable price, potential sellers (the present owners) are neither inclined to increase nor reduce their holdings by more or less than potential buyers are inclined to change theirs. The important natural good for our purposes is land.

On the other hand the prices for *produced goods* depend on unit costs of production plus a gross profit margin. Unit costs of production are direct costs, consisting of unit labour and material costs, plus unit plant cost (if plant is hired). Gross profit margin covers the unit contribution to overhead or fixed costs, plus net profit per unit. If unit costs of production are constant with respect to volume of output, then a change in demand will affect volume produced and sold, but not the unit direct cost. If, however, unit direct costs increase with volume, then a demand increase pushes up unit direct cost, leading to an increase in price.

In the short run, an unexpected increase in demand for a produced good may raise prices well above the unit direct cost plus normal mark-up level, but this is because it creates a purely temporary excess of demand over *engineered* capacity (the maximum output present establishments are capable of producing, even accepting increasing unit direct costs). In a fixprice market stocks would be run down, and waiting lists or delays would arise as well as price increases. In the longer run, either total capacity to produce the good in question would be expanded, as existing producers invested in extra capacity and new entrants set up or introduced new capacity, or the increase in demand will prove temporary, perhaps by inducing buyers to switch to substitutes to escape the shortage prices or delays.

Similarly, an unexpected drop in demand may cause prices to fall well below unit direct cost plus normal mark-up level, by inducing a panic amongst competing producers each trying to keep their capacity relatively fully utilised, in conditions of high overall excess of capacity over demand. Again, though, this will be short-lived, as producers soon react by scrapping or disposing of much of their excess capacity. In a fixprice market, stock or inventory levels would rise to begin with, followed by discounting of stock and price reductions.

We assume that a certain amount of spare or excess capacity is the normal or usual condition experienced by most industries and firms. That is, each firm could produce significantly more output than it is doing, without encountering rising unit direct costs. In the USA, industrial capacity utilisation rates are regularly measured and reported. There the *norm* appears to be around 80 per cent. Therefore only exceptionally large unforeseen demand increases create shortage conditions and prices, when increases in demand are not predicted in time to bring investments in new capacity on-stream. In some industries, like electricity generation, the period required can be as long as ten years, whereas in others, like quarrying or brick production, the period is much shorter. We will see below that while it is hard to apply this normal concept of capacity utilisation to a firm in the construction industry, it is easier to apply it to the construction industry as a whole. At the same time the concept works admirably for most building materials producers.

Nonetheless, changes in the level of demand may affect the size of the gross profit margin, in construction as in other industries. When demand increases, competition between suppliers eases and a sellers' market appears. Firms find themselves able to raise their margins without losing competitiveness relative to rivals, who are also raising their margins. Sellers' markets are more commonplace in construction than in most manufacturing, precisely because construction firms do not carry normal spare capacity in the same way as manufacturers. Moreover, as output units in construction are buildings and structures, they are usually one-off designs and not easily comparable, so construction services tend to be more mixprice than fixprice.

Capacity at the level of an industry or sector

Hillebrandt (1984) defines capacity of an industry in any time period as the maximum output attainable within the limits of acceptable strain. This definition of industrial capacity corresponds to the idea of *engineered capacity* at the level of the firm, but with social factors taken into account. Capacity is not a static or constant amount. It is a potential volume of production, which can vary over a period of time depending on the expectations of key actors in the industry's production process. The main such actors are firms. For firms it is expectations regarding sales, and actions deriving from those expectations, such as investment in fixed capital, in training or in research and development, that raise industry capacity.

However, in addition to the activity of firms, industry capacity is profoundly affected by investment decisions by both government and workers, and allocation decisions by workers, who decide where to work and what skills to learn or practice. Workers' decisions to commit themselves to working in one particular industry rather than another play a crucial part in determining the total labour force available to that industry, and hence its capacity. The individual members of the workforce of an industry make decisions about how much to *invest* in themselves, as what economists now call *human capital*.

Government invests in raising an industry's capacity when it commits resources to expanding education and training provision geared to that industry, or invests in research and development relevant to an industry, or in basic social and physical infrastructure that will raise the capacity of each firm that can utilise it.

The point is that an industry is not an economic actor, capable of making economic decisions to increase capacity. Capacity depends on the reservoir of resources of labour, fixed capital, and technology, and not, directly, on firms. Thus, for example, an increase in the number of firms does not, of itself, add to the capacity of an industry, but only to the number of firms competing to use its resources. Nevertheless, it is firms that can cause increases in the productive resources available to an industry by acts of real investment in the broad sense.

A firm can be said to own a certain capacity with respect to the capacity of its fixed capital stock and hardware-embodied technology. But we must be careful about attributing to the firm the capacity of its employees. They are not *owned* by the firm, and are free to move between firms. If they do, capacity of the industry is shifted between firms, but neither lost or gained necessarily in aggregate. Whether or not there is such a gain or loss in industry capacity resulting from the shifting of productive resources between firms depends on the net effect of two factors that will usually pull in opposite directions:

- the temporary loss of social or team-based productivity that will always result when an individual producer moves from a team and institution into which they were integrated and with which they were familiar, into a new context where learning-by-experience and team-building will have to begin again; and
- possible differences in efficiency between firms, based chiefly in the possession by one firm of superior technical knowledge.

In the long run, firms may enter or exit the industry, thereby changing its capacity if they invest in or scrap their own capacity.

We define economic capacity as the volume each firm would like to produce, in order to minimise its costs per unit. In the short run we are interested in the point at which unit direct costs rise to equal or exceed unit selling price, for this becomes the upper limit to output beyond which the profit-seeking firm will not go.

Since the 1930s it has been observed by all those who have systematically asked firms about it, that most firms in most industries most of the time work well below their capacity level, when capacity is defined as the point at which unit *total* costs would start to rise significantly. The reason for this is that firms in most industries are driven, by oligopolistic competition and the pursuit of economies of scale, to invest in larger and more plants than they can fully utilise most of the time. Each firm is reluctant to scrap its own spare capacity. Each wishes to have enough capacity to meet its more optimistic estimates of sales, and to meet any short lived peaks in demand. Moreover, the result is not a lack of profits, because the industry price level adjusts upwards to equal normal capacity utilisation levels of costs, plus a normal profit mark-up. If sales do increase above this normal utilisation level, then total costs per unit fall, as the high fixed cost lump sum can be spread over a larger number of units of output, and profit per unit sold increases even if prices stay unchanged.

This picture would apply, for instance, to most of the industries manufacturing basic construction materials, like cement, bricks, steel and glass. However it does not fit either construction industry firms assembling structures on site, or firms providing construction professional services. In construction we find no equivalently large fixed investment costs. A falling curve of fixed capital cost per unit, though it may exist in the building industry, will not have a significant effect on the curve of total cost per unit.

The neo-classical versus the 'constant unit direct cost' model of supply response to demand

In any given industry the responsiveness of firms in aggregate to changes in price is called the industry price elasticity of supply (PES). An increase in demand increases prices, but these higher prices also call forth some increase in total industry supply, the size of the actual increase depending on the PES. This supply increase reduces the amount by which prices must increase to restore an equality between demand and supply.

In neo-classical theory each firm is assumed to offer supply equal to the volume at which its marginal cost (MC) is equal to price, and at any given price, the volume of each firm's output is determined by the tendency of MC curves to slope upward. It follows that only an increase in price or a downward shift in the MC curve, arising for some independent reason, such as a new technology, can persuade firms to produce more.

However, if unit direct costs are roughly constant with respect to volume, until a volume much larger than current output is reached, then the output and supply level of each firm becomes indeterminate. The firm would wish to produce more than it is doing, and is constrained only by the lack of demand. If demand increases, lifting this constraint, the firm may produce more, without requiring an increase in the price level. However, if all firms simultaneously attempt to increase their production, each may find that its unit costs do actually increase, because each is relying on being able to find and employ spare, unutilised resources from the reservoir that constitutes the industry's unutilised resources at that moment. As the level in the reservoir falls, firms find themselves competing with one another to obtain now scarce spare resources, and in the process bid up the prices of these resources. This is quite different from the case where the firms actually already possess or employ all the resources they need to increase output, when supply probably will increase in response to demand without an increase in price.

In practice, the ability of the construction industry to respond to demand depends on the type and nature of the construction work required, the materials needed, the work force skills required and the complexity of planning permission and design work involved. Extra demand for one type of construction output can usually most easily be supplied by shifting resources to its production from that of other types of construction output. However, there are some resources the bulk of which is always used in the production of one type of output. For instance, most bricks and most bricklayers are used in new house building. If demand for new housing increases, therefore, little extra bricklaying capacity can be found by shifting bricklayers from other types of construction work (NEDC, 1978).

In the long run supply tends to be more elastic than in the short term. However, in the long run factors other than selling price change, and it is impossible to disentangle their effects, for example to distinguish long run elasticity of supply from *shifts* in supply conditions. Therefore, it is probably a good idea to replace the unrealistic distinction between movement along (supposedly instantaneous) and movement between supply curves with a model which instead measures supply-responses in real time. In this new model supply can respond to any measure or indicator of demand, not necessarily just to a price change. We have thus shifted from analytic time to real or perspective time.

In the short term, which Hillebrandt (1984) thinks in practice in construction might be approximately up to three months in real time, existing stockpiles of materials and components can be used and more materials can be imported from stocks in other countries, along established trade channels. Unemployed workers can be given work immediately and overtime given to the existing work force. The plant utilisation rate can then be increased. Finance may be obtained through bank overdrafts and extended credit periods from builders' merchants. In the downward direction, workers may be laid off, supply contracts and orders not renewed, plant not rehired – generally, buyers working on short-term purchasing or hiring contracts can pass the problem of excess capacity on to their suppliers and workers.

Production can be expanded in the medium term, which may translate in real time into a period of more than three months but less than one year, by making minor acquisitions of plant and by re-employing machinery that has been mothballed. With medium term expansion, there is sufficient time to organise new sources or channels of supply and imports, and to attract labour from other industries, or from abroad or from retirement, especially those who may have previously worked in construction. When the change in demand is downward, fixed capital utilisation rates fall, and some highly skilled, mainly professional and managerial labour is underutilised but still employed. Retiring resources, whether people or machines, are not replaced. Overhead elements of production plans are strategically reviewed and some are cut.

In the long term, which may translate in real time into more than one year, expansion requires strategic investment, available innovations to be implemented, and training, but this will only be done if firms see that a sustained increase in demand is likely. Over the very long term of five years or more, a search for new technologies and innovations can be initiated which, when developed, increase capacity. Moreover, in so far as industry capacity is constrained by limits to the willingness of *existing* firms to invest and expand, in the long run *new* firms are constantly being formed. However, as noted earlier, more firms only means more capacity for the industry if those additional firms take steps to create their own resources rather than just compete for a given pool of resources with existing firms. In the long run, as Hillebrandt (1984) points out, resources have to be created, in contrast to the short and medium term when use has to be made of existing resources.

Unit labour costs over the business cycle

In addition to the effect of the level of demand on the level of wages per unit of labour time, there may be a further effect on costs via an effect on labour productivity. Unit labour costs reflect both wages per hour and productivity, and are measured as labour cost per £ unit of output.

When general demand for labour is very strong, shortages of particular kinds of skilled labour may force firms into inefficient substitution by less-skilled labour. Put another way, the extra workers recruited at the margin to deal with exceptionally high demand may be less productive than the better workers that firms selected first.

As a quite separate effect, the productivity of the same personnel may fall in boom conditions. If workers know that they can easily find other jobs at similar or higher wages, the cost of job loss to the worker is greatly reduced. If employers rely on the ultimate sanction of dismissal to raise workers' work effort, they may find that the effort forthcoming per hour or per £ of wages paid falls in a full-employment boom (Bowles and Edwards, 1993).

Again, peak levels of output may be achieved by heavy use of overtime working. The effect of overtime on unit labour cost in the construction sector varies dramatically between branches. In some professional service firms, the internal ethos is such that firms can persuade permanent professional staff to work unpaid overtime hours. Indeed employment contracts may not even specify the number of hours constituting the normal month for which the basic salary is paid. Such staff may merely be promised lump sum, end of year bonuses for 'helping out'. In contrast, in highly impersonal and casualised labour markets, such as apply for most construction industry operatives, overtime work will have to be paid for at a premium wage rate per hour, even though the productivity of those overtime hours is only at or below average productivity.

On the other hand, if workers identify themselves with their firm, and wish to remain with that firm for as long as possible, there may be positive morale effects of high order books and expansion of the business, as jobs are made more secure and promotion prospects increase.

For the construction industry the evidence on the overall net effect of demand conditions on productivity is unclear and capable of alternate explanations. There does appear to be a tendency for productivity to drop after several years of a sustained boom. However the evidence is that productivity then continues to fall in the early part of a recession (output falling faster than employment). Productivity seems to rise fastest in the latter part of recessions, and in the early part of recoveries.

There are problems which arise in the interpretation of output and employment data time series in this way. The employment series measures numbers of workers employed in the industry, not number of hours worked by all workers per year. During a recession, it is highly likely that the number of hours worked per labour only subcontracted worker per year falls sharply, which is equivalent to a rise in under-employment, though we have no way of measuring this. Output per hour might then be rising even while output per worker per year was falling.

Another problem of any long run time series, due to rising prices over time, is the conversion of output values for different years into comparable terms. To obtain a measure of real change over time, we need to apply accurate price-deflators. This deflated or constant price series for value of output we then use as a proxy for volume of output. However, the actual deflators used may either under or over-estimate the proportion of the change in current-price value of output from one year to the next that is attributable to price changes rather than real volume changes. One crude but effective check is to compare the constant-price volume series with physical unit series for key construction inputs, like tons of cement or thousands of bricks. These inputs may be calculated using official data for national production less exports plus imports (DETR, *Housing and Contruction Statistics* and *Monthly Statistics of Building Materials and Components*). If physical series show a rate of change that is very different from that of supposed volume of output series, it is at least possible that the reason lies in an error in the latter. This check works less well if the composition of construction output between output types is simultaneously shifting.

Measuring spare capacity in the construction industry

The concept of *capacity* is important for policy makers and firms. The capacity of the UK built environment production sector largely determines the ability of the construction economy to meet the demands placed upon it. Ive (1983) has argued that although markets need not coincide with industries in their boundaries, there should be an attempt to relate the two concepts. In practice, however, analysis of construction demand tends to separate markets by region, by type of output, such as housing, offices, and factories, and by type of client, the public or private sector. The analysis of supply meanwhile classifies firms by trade of principal activity, by region of location of the firm, and by size of firm.

There are three empirical methods available in principle to measure capacity in the construction industry. The first looks at peaks in construction output in the past and takes a line drawn to connect the series of peaks in construction to represent the growth in capacity of the industry. The underlying assumption is that at each peak something approaching full-capacity at the time was reached. Obviously, 'engineered' capacity must have been at least equal to maximum output volume actually achieved, but we cannot be sure that even at the peak there was not some spare capacity remaining. Further tests for this would include looking at the movements in input-prices and labour productivity during those past peak periods (if they represented full-capacity, one would expect to find rapidly rising prices and falling productivity).

Another method is to measure under-utilised capacity in industries producing plant and materials. This approach was adopted by NEDC in two well known reports published in the 1970s, *How Flexible is Construction?* and *Construction into the Early 1980s*. One advantage of this method is that it may be these backward-linked industries that constitute the first effective or operational constraint on capacity to produce final construction output. Moreover, it may be easier in practice to obtain *volume* estimates of capacity for such manufacturing industries, based perhaps on measures of installed *engineers' capacity*. This approach requires estimates from many fewer but large firms, each with their own resources, with the benefit of reducing problems of non-additivity of different types of inputs.

The third method of estimating construction sector capacity would in principle be to measure unemployment, or perhaps unemployment-to-vacancy ratios, for each construction sector industry where labour is believed to constitute the operational resource constraint. However, since 1981 UK unemployment statistics have not been broken down by industry, though the regular Labour Force Surveys (small-sample censuses of the economically active population) do yield some information on the industries in which people currently unemployed are looking for work.

Whichever measure is used, it is clear that there has been chronic and substantial spare capacity in UK construction, in all bar a few peak years. What is not clear from such methods of measurement, is the extent to which former capacity is lost during each recession, either by the scrapping of plant and equipment, or by the exit of workers from the sector, or by loss of skill levels from disuse.

From 1981 to 1990 construction output supposedly increased by 59 per cent in volume terms. This conclusion can be derived from the value of output at constant 1990 prices given in Table 1.6c in *Housing and Construction Statistics, 1983–93* (1994). Recession then led to a dramatic decline in demand for construction: orders for new construction obtained by contractors, at 1990 constant prices, fell by 27 per cent from 1988 to 1992 (ibid., Table 1.1c); output of all work at constant

1990 prices fell by 12 per cent between 1990 and 1993 (ibid., Table 1.6c).

Imports and construction industry capacity

Construction is a national rather than an international market. National demand is largely met by national supply. Unlike other industries, imports and exports in construction are marginal. According to *Housing and Construction Statistics, 1983–93*, Table 4.6, total imports of building materials in 1989, at the peak of the last boom, had a value of £4,700 million compared with a gross value of construction output in that year of £52,150 million. Imports represented around 9 per cent of total required capacity.

The ability and capacity of the construction sector to meet the demands placed upon it by the economy can be measured, in the case of materials manufacture, in terms of import penetration. For the construction industry itself, lack of capacity might show up in inward labour-migration, in relocation of projects outside the UK, in lower quality of construction (substitution of quantity for quality), delays and longer lags to the construction process, and in much higher construction prices during booms and near peaks.

There are a variety of reasons for importing construction materials and services. Hillebrandt (1984) quotes five groups of imports listed by the Building and Civil Engineering Economic Development Committees (EDCs) as differing in the reason for their importation. They are:

- Wood and other renewable or non-renewable raw materials that are unobtainable, for natural/geographical reasons expressed in 'intrinsic' comparative disadvantage to UK production, in required quantities and qualities in the UK. In the early 1980s, these accounted for nearly half the value of all imports.
- Items temporarily in short supply often due to an unexpected rise in demand. In the later stages of the commercial and housing property boom of the 1980s, brick production lagged behind demand making it necessary for contractors to secure supplies from abroad.
- Materials in which a foreign supplier has become established as a secondary source of supply, often originally in a previous shortage.
- Products regarded as superior in some way to those available domestically, for which no direct UK substitute of equivalent quality exists. This was the case with curtain walling which was

imported from Germany, for instance, to meet the requirements of the developers in the Broadgate scheme at Liverpool Street Station in London, and Canary Wharf in Docklands, both projects seeing quality as a major selling point of their property developments.
* Products that are cheaper and/or of lower quality than those available in Great Britain.

Capacity and international construction

Until very recently, it was appropriate to analyse UK construction entirely as part of a national economy. Demand and capacity were both essentially national. International competitiveness was of no direct significance. Construction prices in one country bore no necessary relationship to those in another. This has all begun to change and the process of change has profound implications for construction economics. Some of these implications are dealt with here.

The capacity of any industry needs to be related to demand. The ability of the national construction industry to satisfy the demand for new build, refurbishment, repair and maintenance work, depends on its capacity or maximum potential to produce output in any given period. This in turn depends on the total of both employed and unemployed resources of skilled and unskilled labour, plant and equipment, materials and management. Where capacity is insufficient, the economy will import but when there is excess capacity, firms will tend to seek markets abroad. Hence, trade and capacity are related.

Engineering News Record (ENR) data shows the existence of a vast market in international contracting and design. However, inspection of the ENR data reveals that the great majority of this international trade comprises exports from a small number of countries with highly technologically advanced construction industries compared to the rest of the world, especially to developing or less-developed economies. There is relatively little trade between countries at broadly similar levels of economic development, and relatively little importing of contracting or design by countries with advanced industrial economies (Campagnac, Winch and Lin; Linder, 1994, 2000).

Housing and Construction Statistics, Table 1.11 (b), shows the value of work done abroad. Because this is shown at current prices, it understates the decline since 1984–5, just as it over-states the earlier growth. It is important to realise that much of the value reported here concerns the output of overseas companies acquired as subsidiaries by UK parent groups, as part of their strategy of internationalisation of their assets, rather than actual exporting by UK-based operating businesses.

Towards 'single European markets' for construction and for building materials?

Building industries and building products vary from country to country. Laws and legal systems vary. Resources, materials and skills vary. Building requirements vary depending on climatic conditions. Costs of materials vary. Acceptable standards vary. There are also cultural variations.

The division and combination of roles and the types of participants (actors) vary from country to country. While subcontracting is common to all countries, it varies in degree from one to another. In Germany firms tend to own their plant and equipment, hiring only as a last resort. In the UK there is a thriving plant hire sector. However, nothing is permanent and there is a rapidly growing plant hire market in Germany. Even within Great Britain practice varies. For example, in Scotland the role of professionals is at variance with English practice, as is the law.

Recent European Union (EU) Directives have sought to create the institutional basis for a single European market in construction, especially in construction materials and components. National regimes of building regulations, safety codes, etc. are supposed to be replaced by European standards; so that products approved for use in any one EU country will be deemed thereby to satisfy legal requirements in any other EU country.

However, each national construction industry remains largely idiosyncratic and highly regulated and shaped by its own national systems, whether contracting systems or other institutions, so that while the industry exists globally, it varies from country to country (Winch, 1996). Ball (1988) has pointed out that one of the peculiarities of the construction industry is that these differences between countries have prevented foreign competition from developing in any of the advanced capitalist economies.

In this respect the construction industry reflects the idea of institutionalist economics that economics is not a universal science but one which can be used only to describe specific sets of social, legal and economic conditions.

Individual actors, accustomed to playing one specific role within the context of a national contracting system, find it very hard to internationalise that role. Their source of apparent international competitive advantage may disappear, by the time they adjust their role to accommodate to the requirements of participation in a foreign system (Winch and Campagnac, 1995).

Concluding remarks

In this chapter we have seen that capacity in the construction industry refers to the resources available for the provision of built environment. These resources include labour, plant and materials. The firms in the construction industry mainly merely utilise existing resources, through sometimes making investment decisions which increase capacity of the industry by, for instance, purchasing new plant and machinery. These plant and machinery investment decisions are more usually taken by plant hire firms, rather than directly by contractors.

We find Hillebrandt's (1984) definition of construction capacity, as the maximum potential output within acceptable strain, sufficiently flexible to take social factors into account. This is important as capacity is constantly changing in response to changes in demand. Finally we recognise that capacity is influenced by international factors such as the international movement of labour and materials.

8
Capacity of the Firm and the Economics of the Growth of Firms

Introduction

In the preceding chapter we discussed the meaning of capacity for industries in general and for construction sector industries in particular. The meaning of excess capacity in *firms* in the building materials industry is clear, and refers to under-utilisation of capital intensive plant. Its meaning and existence in construction industry firms is less clear. If it exists, it will be largely a matter of under-utilisation of the firm's core staff. 'For the most part, "excess capacity" is externalised by the contracting firm, appearing as unemployment of construction labour, under utilisation of hire firms' plant; increased stocks, or reduced output, in the industries producing building inputs; (money) capital held liquid e.g. cash or short term deposits; and a flow of money-capital from construction to other sectors, channelled through both the banks and multi-industry firms' (Ive, 1983, pp. 72–3)

In this chapter we discuss both modified neo-classical and post-Keynesian approaches to cost curves of firms and describe how firms behave in construction markets. We adapt Penrose's approach to capacity and discuss economies of growth to show how firms are driven to expand in order to remain competitive. We conclude by adopting a behavioural approach to explain company expansion, based on management's ability to diversify and make the firm grow.

The shape of the construction firm's cost curve

The key preliminary point is to be clear about what a cost curve diagram tries to show, and what it cannot be used to show. The x-axis purports to measure changes in the volume of output of the single firm, as a result of its management's decision, relative to an implicit constant level of industry output volume. The y-axis purports to show the effect of such single-firm volume changes on the average level of costs per unit.

Thus we will describe a *firm's* costs per unit as constant with respect to volume, whilst at the same time stressing that the *industry's* costs per unit (consisting of an average of the costs of all firms) vary significantly and directly with the industry's volume of output. In terms of the single firm, the latter effect can only be represented in terms of a shift or jump from one cost curve to another, higher or lower than, and not necessarily parallel to, the old curve.

Unit costs of all firms vary directly with industry output volume, because the latter represents a change in the balance between industry demand and industry capacity. If, say, industry demand rose whilst capacity remained the same, we would expect increases in the unit prices of any inputs supplied via flexprice markets, for example, casually employed labour, spot-market material inputs, or land. In conventional terms this would be the result of a shift in the demand curve relative to an unchanged supply curve. Changes in labour costs per unit, as discussed in Chapter 6 (pp. 179–80), are in fact probably the most important source of this kind of industry-wide cost change.

In addition, if the general rate of inflation in the economy is positively linked to the aggregate level of demand (as it probably is, for the most part), and if changes in construction demand coincide with changes in aggregate demand, then we would expect to find all full cost or fixprice construction input prices changing in line with the level of construction demand. General inflation works its way into the input costs of construction materials and component industries. These cost increases are then passed on by these industries as higher prices charged to the construction industry. Thus, all construction firms experience a similar upward shift in their unit costs at the same time.

The neo-classical rising short-run marginal cost curve

Let us attempt to apply an orthodox neo-classical model. The *marginal cost* model requires that each firm eventually encounters diminishing returns (increasing marginal costs per unit of output). Otherwise each firm's profit-maximising output would be infinitely large. The law of diminishing returns assumes at least one input remains the same. As other variable inputs are then increased, eventually returns from successive units of inputs decline. In that case the question is, what fixed inputs in a construction firm are responsible for diminishing returns in the short run? Several authorities on construction (e.g. Hillebrandt) have proposed management resource rather than fixed capital equipment as the key input in this respect (taking a lead from Penrose's theory of the limits on the rate of growth of the firm), because they assume the existence of plant hire and building materials industries with adequate spare capacity to accommodate any growth in planned contractor output in the short run, and of a casual labour market with significant numbers of skilled workers under- or un-employed, so that the construction firm can readily increase its use of all other inputs besides management.

Alternatively, one could explain rising costs not in terms of less efficient production (a worsening in the technical ratios at which input quantities are converted into output quantities) for reasons internal to the firm, but rather as the result of rising prices for input resources and labour, resulting from the fact that all contractors are simultaneously attempting to increase output, and thus raising aggregate demand for construction resources. This would involve a shift in, and not a movement along, the cost curve. Of course, such increase in input-prices, being general, might well eventually be reflected in a higher market price for construction output, but this would take time to occur; it might therefore set an interim upper limit to the firm's output volume.

It seems at least possible that initially unit costs would not rise significantly as industry output began to increase, but that after some time had passed there would then be a rather strong impact on costs.

However, in what follows, rather than the neo-classical concepts of marginal and average cost, we will distinguish between direct and fixed or overhead costs, rather in the manner of firms' cost accountants. Also, instead of the conventional neoclassical distinction between the short and the long run in terms of fixed and variable inputs in the short run and only variable inputs in the long run, the short run we will redefine in managerial terms as involving decisions not referred to central management, while long-run policies and decisions are defined as those only resolvable by central management.

The mirror-L-shaped unit cost curve

One common assumption is that the most realistic way to think of unit direct cost curves is as mirror images of an L-shape, that is, constant up to a point, but thereafter rising vertically, or nearly so, so that no more output can be produced except with prohibitive increases in unit costs. The x-axis correlate of the point of kink in the slopes of the cost curve then clearly represents 'capacity'.

We broadly concur, but would put the matter in *uncertainty-plan* terms. It does not matter for our argument that many firms do not think of themselves as possessing such plans. In effect firms that do not consciously plan for contingencies are simply assuming that the near future will be within a certain range of the recent past. Such firms, too, therefore have actual, if unarticulated, upper and lower limits within which they can produce without costs rising out of control. The firm thus has a production plan built around upper and lower scenarios of what is sufficiently possible to be worth anticipating and a contingency plan to meet either scenario if necessary (this, we believe, corresponds to Shackle's (1952) idea of 'degree of potential surprise'). Within this contingency range, unit direct costs will be more or less constant. If, however, the firm attempts to operate above the maximum output considered when forming its production plan, then it will encounter very steeply rising unit direct costs. We believe this interpretation is fully consistent with the interview responses of construction firms in surveys touching on 'capacity' by Hillebrandt *et al.* (1995) and Hillebrandt and Cannon (1990).

Firms engaging in 'full cost' or 'cost plus' pricing calculate their overhead or fixed cost per unit of output by dividing planned output (say, a volume somewhere in the middle of the expected possible range) into total (known) fixed costs. This fixed cost per unit then becomes the *contribution* to total fixed costs required from each unit of output. Price comprises unit direct cost plus contribution plus profit margin. The two latter together comprise the gross mark-up. When a firm finds itself with the possibility of exceeding its originally expected output volume (because demand is unexpectedly high), in cost terms it encounters two phenomena pushing in opposite directions. On the one hand, it will by that point have achieved sufficient 'contributions' to cover its total overhead or fixed costs (comprising central administration costs, fixed capital costs, and costs of core project staff). On the other hand, it will at that point encounter steeply rising direct costs per unit, particularly (though not only) because the costs of staffing such additional projects with managers and professionals shifts from being a part of overhead cost to being a direct cost, as new (temporary) staff must be hired or obtained via subcontracts with other firms.

Staff and capacity

It is clearly important that we allow in our theory for the fact that construction firms do have, and make use of, relatively easy recourse to subcontractors and to temporary workers. Thus we cannot simply draw their *capacity* as an engineered-in vertical line. On the other hand it is also true that most firms have a core of project-based staff, whose numbers they take as given in the short run. These might be site or project management staff in a management contractor, or project architects in an architectural firm. One meaning of *full capacity* is, then, the range between the minimum and maximum level of output associated with efficient deployment of this core of directly productive staff.

Suppose these core staff are working at full capacity, for the relevant future, on projects already won and in the order book. How then will the firm behave in relation to taking on further projects? Construction firms do not like to turn away invitations to tender, unless they are so busy that internal organisation is becoming clearly chaotic. However, in the situation described they may choose between two strategies. They may raise their tender prices, to reflect the extra costs, uncertainties and difficulties of having to take on and manage temporary staff should they win the bid. Alternatively they may cut their prices, to reflect the fact that they have already reached their output target for the period, and already fully covered their overheads for the period, so that they could now take on work on a 'direct (probably subcontracted) cost plus profit margin' basis, without contribution to overhead recovery.

The latter would imply that we would find prices falling in times of strong demand, when firms can reach full staff capacity levels of orders. Conversely, presumably, if firms' order books had fallen short of their target, we would then have to expect to see them raising their margins, so as to recover their overheads from a smaller volume of projects. Now, monopolies have been observed sometimes to price in this way. When demand is buoyant and profits are rolling in, a utility company, say, may offer customers a price cut. However, the aim of such price cuts is not then only to increase demand, but rather also to placate and forestall possible regulation, windfall-profit taxation or anti-trust measures.

The reasons why most construction firms do not price work based on a direct subcontracted cost plus normal profit margin at full staff capacity may be because:

- most construction firms only find themselves at full capacity when the whole industry is also at or near to full capacity. In these circumstances it may be impossible to find staff-function subcontractors at prices that would be profitable. Also, at such times it will be easier and probably more profitable instead to raise gross margins until excess demand is choked off by the price increase.

- Some construction firms may be guarding their reputation for quality, or their legal liability for defective service. This is obviously put at risk by contracting-out key staff-functions to unknown and temporary subcontractors in times of peak demand.
- Most construction firms will be guarding their ability to control out-turn unit costs. When margins are small, as they are usually in construction, then even an apparently relatively high profit margin may be more than eaten-up by uncontrolled direct project costs, which might result from use of poor or poorly integrated temporary staff to control projects.

On the other hand, the boom of the late 1980s did see many staff-intensive construction firms (design firms as well as construction managers) achieving a rapid increase in order books by heavy use of temporary staff even for key functions on projects – so we must be careful not to overstate the reasons why this 'will not happen'!

Evidently, to find itself operating at beyond full capacity, the firm must be suffering from bounded rationality of decision making. If it is more concerned to guard its reputation than win additional work which generates a net profit, it cannot be a short-run profit maximiser by objective. The perfectly informed profit maximiser would have foreseen the increase in demand, and raised its prices from the beginning of the period to that level which would choke-back orders over the period to just equal maximum staff capacity. It would charge the maximum the market would bear, for its finite upper level of volume of sales and output for the period.

We think that the balance of the evidence supports the view that, having set a system of gross mark-ups (contribution plus profit) for a time period, firms do not cut them simply because, as things have turned out, they can afford to do so. Rather, variable mark-ups, which do exist, arise because firms identify their market as comprising competitively harder and softer segments, but this price discrimination operates simultaneously rather than sequentially.

In effect most firms think in terms of a single composite mark-up on direct cost, to cover both overhead and profit. However, they do not break this down into its components at project level, precisely because such a breakdown only makes sense *ex post* at the year's end. Only then can actual volume be known, and therefore the overhead cost per unit and profit per unit.

A Penrosian idea of capacity applied to a construction firm

Edith Penrose, in *The Theory of the Growth of the Firm* (1st edn, 1959), made two fundamental contributions to the theory of capacity. She was the first to conceive the capacity of a firm primarily in terms of its management rather than its stock of fixed capital equipment, and thereby to form a mental bridge between the economic theory of capacity and management and organisation theory. She argued that it was easier for a firm to increase its stock of fixed capital, in any relevant period of real time, than it was for the firm to increase its effective management capability. Thus the firm's 'stock of management capability' becomes the key or operational constraint on the firm's ability to raise its output through time. This was her first contribution.

Her second contribution was to re-conceive *capacity* in purely dynamic and relative terms. Conventionally, the optimum capacity of a firm was assumed to be the output corresponding to whatever stock of fixed capital would minimise its long-run total average costs, mainly considering optimum long-run labour/capital rates of substitution and economies of scale. Full utilisation of such a stock then constituted full capacity operation. This full capacity was an absolute rather than a relative magnitude. Instead of this way of thinking, Penrose proposed that firms set their target capacity (maximum feasible output volume) for next year in relation to their actual present capacity. More specifically, she argued that they set no absolute upper limit to their ultimate capacity (no long-run *optimum* size), but rather set an annual limit to the rate of growth of their capacity from its present level.

Raising a firm's capacity through time, for Penrose, was first and foremost a matter of recruiting additional management staff, integrating those new staff into the organisation's teams, allowing the passage of time to give the experience of working together that alone can create effective teams, so that the new staff became as effective as existing staff, and restructuring that organisation if necessary, to adapt it to a change of scale of operation.

Growth by acquisition is not featured strongly for two reasons. First, any acquired firm would bring with it its own customers and demand as well as its managerial capacity. Second, acquisition greatly raises the amount of scarce managerial time that will have to be devoted to integrating and restructuring the two businesses, and thereby, at first, tightens rather than loosens the managerial capacity constraint on operational efficiency in each of the businesses considered separately.

Only after restructuring and integration are complete are spare resources freed up.

Penrosian firms diversify, so as to free themselves of the constraint on growth of demand in their initial product market alone, but they do so from below. The diverse product businesses are managerially integrated, and not merely parts of a financial holding company.

Each firm would discover, from its own experience, the maximum rate at which it could increase its staff establishment without encountering loss of managerial effectiveness or efficiency. Let us denote this rate as 'e'. Suppose we 'pick up' the continuing story of a firm at a point where it has some spare managerial capacity, presumably because demand growth has been less than expected. Suppose that demand now starts to grow faster than before. As demand for the firm's output now grows through time, the firm will first of all utilise its existing spare capacity. This is its short-run response. If it projects the new rate of demand growth, 'g', to be likely to be long-lasting, it will embark on a long term plan of expansion of managerial capacity.

If g is greater than e, then it is the value of e that will determine the rate of growth of the firm. If g is less than e, then the growth rate of capacity will actually be set by the value of g, in which case, the possibility of diversification will come into play. The *cost* of raising the constraining value is either: the extra cost of diversification (research and development, and initial market penetration and entry costs; lack of economies of learning or experience), if g is the operational constraint; or, the extra cost of excessive rate of organisational expansion, if e is the key constraint. Eventually, if the actual rate of growth of output is pushed too high, one or both of these additional costs will become sufficiently severe that profits will fall, and the firm will adjust to a more modest growth rate.

Hillebrandt *et al.* (1995) adopted a Penrosian approach to enquire into growth and capacity of construction firms. They found that many chief executives not only agreed with the idea that their size was only constrained by an upper limit to the efficient rate of annual growth, for managerial efficiency reasons, but also that they themselves thought about the issue in more or less these terms. Whether maximum growth rates are specified in pure proportional terms, or as a rising stream of annual target turnovers, and whether in constant or current price terms, may seem minor technicalities. However, each would result in quite different properties to the model, when formalised.

For some specialist construction firms, we here propose an extension of the Penrose model to include specialised skilled productive workers as well as managers in the *limited* internal resources of the firm. This might apply, for instance, to a firm of architects. There are two arguments behind this. First, from the point of view of the firm using team-working to produce and sell a highly specialised product or service, it is as difficult and time-consuming to add *core* producers to the firm as to add management. The greater the degree of implicit knowledge and idiosyncracy of the firm's 'way of doing things', the greater the force of this point.

Second, when productive staff are recruited from outside the organisation, they bring with them only a potential to operate effectively within the firm, rather than a fully portable productive capacity. The full development of this productive potential depends upon their integration into the culture, practices and teams of the firm. This is the opposite assumption to that of a system of qualification or professional certification of skill, in which the individual is certified as *bearer* of productive skills and capacities assumed to be fixed and independent of the context or firm in which they are exercised.

Penrose argues that firms' controllers' motives are best described as desiring the growth and survival of their firm.

> Individuals thereby gain prestige, personal satisfaction in the successful growth of the firm with which they are connected, more responsible and better paid positions, and wider scope for their ambitions and abilities. (Penrose, 1980, p. 28)

The controllers therefore have 'a desire to increase total long-run profits' (ibid., p. 29); to grow by means of investment so long as the return on investment is positive; and therefore a desire to make and retain as much profit as possible, because retained profits are the means of financing successful growth.

> If profits are a condition of successful growth, but profits are sought primarily for the sake of the firm, that is, to reinvest in the firm rather than to reimburse owners for the use of their capital...., then, from the point of view of investment policy, growth and profit become equivalent as the criteria for the selection of investment programmes......to increase long run profits of the enterprise in the sense discussed here is therefore equivalent to increasing the long run rate of growth. (Ibid., 1980, p. 30)

However, this is not just growth for the sake of growth. Firms need to expand in order to survive, as the following argument illustrates.

Penrose discusses the existence of *interstices* or gaps. An interstice arises out of spare capacity. Spare capacity occurs whenever the volume of production fails to utilise all plant and labour. As individual items of plant and particular skilled or specialised employees are invariably used at less than their full capacity, there are always a number of interstices.

This can be explained simply. Assume three types of machines, A, B and C. The capacity of A is 3 units per hour, of B, 4 units per hour and of C, 5 units per hour. The minimum output that would utilise all machines would therefore be an output of 60 units per hour. The number of machines required would be 20 machines of type A, 15 of type B and 12 of type C. Any output of less than 60 units per hour would mean that some plant would be used at less than its full capacity. For example, if output were restricted to 12 units per hour, then the firm would need 4 type A machines, 3 type B machines and 3 type C machines. However, the full capacity of the type C machinery is 15 units per hour. In other words, 3 units of output per hour are not required and the machines are only used for 80 per cent of their capacity. Interstices arise because it would be a coincidence if the market for a firm's output matched its full plant and labour utilisation.

Applied to labour, interstices arise when specialist staff are not fully employed, and are found other work to do in the meantime. This applies equally in construction, where a contractor's core staff may not always be used to carry out their particular skills, and instead are asked to carry out work of a more general nature, with a lower marginal product than their expertise could generate. A site manager with no site might be required to carry out clerical work in the head office to fill in time. This arises because the volume of work is not determined by the human resources available, but is constrained by the amount of work for which the contractor has successfully tendered.

Several implications arise out of this under-utilisation of capacity. Firstly, underutilised plant and labour represents an additional burden on the cost per unit of output. Spare capacity can be viewed as an addition to fixed cost per unit which is therefore higher than it would be if output were to be increased. If costs per unit are higher than a rival's, the firm is less competitive.

More positively, unused capacity can be used to expand into a new line of output or a new market at no extra cost in terms of the unused capacity. The cost of diversification would only extend to the cost of *new* machinery, and the *extra* labour and resources required. As this would be less than the cost of purchasing *all* inputs required, the savings produce a competitive advantage for the expanding firm over the firm which is not growing. These savings of expansion can be viewed as *economies of growth.*

Economies of growth enable firms to remain competitive in contrast to incurring the additional costs of owning surplus capacity. If firms fail to expand, their total costs per unit make the firms less profitable and the reduction in their profitability reduces their ability to invest in future expansion and diversification. If firms cannot expand they lose their remaining competitiveness and profitability. Consequently, sales decline, turnover is reduced and employment decreases. Finally, companies may become vulnerable to takeover.

Concluding remarks

This chapter has been concerned with the dynamic nature of all firms and in particular construction companies. We have applied Penrose's ideas of management within multi-project firms to show how managements continually seek to expand and diversify the activities of their firms in order to remain competitive and survive. This implies that there is an imperative for firms to grow. As a result they need to invest. As investment relies on both external borrowing and internal funding, we now turn in Chapter 9 to the need to plan for expansion through investment, and the linkages between investment, profits and pricing.

Part IV

Pricing and Investment Strategies

9
Pricing and Investment

Introduction

In this chapter we model the pricing and investment decisions of firms, using the Eichner – Wood model (Eichner, 1980; Wood, 1975). This model will bring together the concepts of pricing, output, finance, profit margins, and gross operating profits, but it is the inclusion of growth which introduces a dynamic element, allowing for change and uncertainty. We shall use the Eichner model to demonstrate the trade-off between the need for internal funding for investment and the constraint of the market on a firm's ability to raise these funds by increasing its pricing mark-up. We adopt Wood's approach to determining the maximum rate of sales growth and profit margin and suggest that this model can be applied to firms working in the construction industry.

We conclude this chapter by looking at the implications of different pricing decisions on gross operating profits and combine the accounting concept of gross margin with the economic concept of price elasticity of demand in a practical application of the Eichner-Wood model.

The determinants of price and long-run capacity of a firm under uncertainty

The post-Keynesian microeconomic literature (Lee, 1999; Arestis, 1992) contains a family of models, in which it is the need for investment finance by the firm that determines the level of profit margin and hence the price that it sets. The models begin by making several assumptions. They assume that the objective of the controllers of firms is growth. They further assume that controllers value their security and independence, and therefore aim to make retained profits sufficiently large to finance this growth. There is a bias against external borrowing. Growth seeking is seen as complementary to profit seeking.

The key difference between the managerially controlled firm and firms controlled by their owners is that the managerically controlled firm will prefer to retain profits rather than distribute them to shareholders. The scope given to this preference may however be limited by threat-of-takeover constraints (Williamson, 1966; Heal and Silberston, 1972).

It is also assumed that industries are essentially national. This model's relevance to the construction sector is increased by its assumption that international competition is not decisive or important for price levels. National firms do not face a price squeeze derived from an inability to pass-on costs or raise prices for fear of loss of international competitiveness.

The model also assumes that the industry structure is oligopolistic. It might seem at first sight that this makes it inapplicable to the construction industry with its low aggregate concentration ratio. However, as we have argued in earlier chapters, the British construction industry now contains an oligopolistic market sector, comprising a limited number of large firms engaged in the supply of management of large projects or in the supply of bundled services for projects. Prices set in the market comprising these suppliers and their clients in turn influence other construction prices.

The basic building materials industries, whose gross output value accounts for a significant proportion of construction gross output value, fit the oligopoly assumption very well. However, in some building materials markets recently there has been an increase in international trade, and a mild tendency to international price convergence.

The model also assumes that industries have firms which are recognised *price leaders*: that is, firms that are completely confident that other firms will follow any price increase they introduce. However, this price leadership assumption is particularly hard to apply to the construction industry, where information on rivals' prices and mark-ups is gleaned only with delay and imperfectly, out of an averaging of data on many individual projects. Instead we propose a response-uncertainty assumption, under which each construction firm is quite uncertain as to how its rivals will respond to a change in its average prices.

The Eichner–Wood model

Prices, and profit margins on unit costs, are set by the firm so as to yield sufficient retained profits to finance the requisite proportion of the firm's long run investment plans. Any general nationwide increase in unit costs will be passed on by the firm as higher prices, whenever it can do so. Similarly, if an increase in taxation or dividends led to a reduction in the ratio of retained profit to gross profit, then the firm would also want to pass the increase on in the form of higher prices.

If investment requirements increase, then firms faced with low price elasticities of demand (in the upward direction of demand curves) deal with this by raising their mark-ups and prices, so as to generate more retained profit or corporate saving to finance this increased investment. This can be seen in the water industry and telecommunications where it has been argued that prices have been set to allow adequate profit levels for investment plans to be fulfilled as well as maintaining dividend payments to shareholders.

On the one hand firms may be assumed to be reluctant to invest, because of the intrinsic uncertainty of returns on investment. On the other hand they see investment as the main weapon of inter-firm competitive rivalry. It is by investing that a firm can gain competitive advantage over rivals and thus increase their market share. If a firm does not invest enough to remain competitive, it loses market share, and in the process the shortfall in its sales and the decline in demand for its services lead to it failing to achieve its aim of growth.

Now, in this context we need to distinguish two broad types of investment. Investment in capacity which adds to a firm's size, and investment in competitive advantage which relates to the quality or relative price of the product or the service being offered. The two are not completely independent, being linked by the existence of economies of scale. Firms can reap a competitive advantage simply by investing in faster capacity expansion than their rivals. In this way a firm can balance growth in demand for its products with growth in its capacity to deliver them. Competitive advantage investment comes first, because it is this that will determine the ratio between growth of demand for the firm and growth of demand for the industry.

Now let

$$G_f = G_i . c_f, \qquad (9.1)$$

where

G_f = the growth of demand for the firm, in *volume* terms
G_i = the growth of demand for the industry, in *volume* terms,

and

c_f = the relative competitive advantage enjoyed by a firm.

If a high level of competitive advantage investment creates a high value for c_f and thus for G_f, then this increases the need for capacity adding investment by the firm. However, the need for capacity adding investment is not solely determined by the value of G_f. It also depends on the existing level of capacity utilisation. Other relevant factors include the search for either economies of scale or economies of growth, similar to those described by Penrose. Also relevant is the rate of retirement or obsolescence of existing capacity yielding assets, such as plant and machinery. If the competitive search for advantage generates more rapid technological change, it may thereby speed the rate of obsolescence of firms' existing capacity, and thus increase the need for investment in additional gross fixed capital formation.

All the above factors therefore influence the level of investment the firm perceives itself *needing* to make. However, there is a trade-off faced by the firm between the rate of investment and demand growth. This trade-off arises insofar as demand is elastic with respect to price. The more elastic the demand the more difficult it is for the firm to raise its price. This therefore sets a limit to each firm's willingness to raise its prices. The more the fall in volume of sales that will result from an increase in price, the less the firm's need for capacity adding investement.

We need to distinguish now between short-run and long-run price elasticity of demand. A firm in an oligopolistic market may face price inelastic demand in the short run. Thus by raising its prices it raises profits in the short run. However, these higher profits in the long run attract new entrants, whilst over time rivals use the price differential to develop marketing strategies which draw away some of the firm's customers, and other customers search out and switch to alternative products altogether. All this describes well what would happen if, say, British Telecom were to raise its prices.

For such a firm, the trade-off is that it can have more profits today, but only at the expense of lower sales growth and lower profits in the future. For construction firms, we suggest the trade-off is still there, but for rather different reasons. Such firms face high price elasticities of demand even in the short run. Therefore their first route to higher short run profits is via price cutting. In the longer run, these cuts lead to lower profits as rivals copy. Their second route is to reduce growth-orientated expenditures in the current period. This will lower current costs whilst leaving current sales untouched.

Beyond a certain point, the monopolistic firm will judge that the *loss* of demand from a price rise in the longer run exceeds the *gain* in increased short term retained profits. Raising a firm's selling prices has a cost. It is assumed that a price increase will increase profits in the short-run, because in the short-run demand is price inelastic. In the longer run, though, more substitution and entry by new rivals occurs, causing the effect on further-distant profit flows to be negative. This implies a link over time between the gains and the losses. From this implied cash flow Eichner derived an implied cost of internal funds, which is the rate of discount that gives the estimated net present value of the changed flow of funds resulting from a price change a value of zero. This is a standard discounting problem and solution, equivalent to finding the internal rate of return.

Wood differs from Eichner in several respects. Usefully, his analysis applies equally to firms that can and cannot raise current profits by raising prices. He makes a sharper distinction between current and target investment levels. Long-run (in practice, 3–5 years) investment requirements are derived from target growth rates for the sales revenue of the firm. The main source of funds to achieve this target investment is again retained profits. Investment plans determine target profits. However, current short-run investment levels are constrained and partially determined by the level or mass of current profits. Thus, the causality between profits and investment in the short-run and between investment and profits in the long-run flows in opposite directions.

The firm faces three constraints on its ability to achieve as fast a growth of sales revenue as it might like. These constraints are demand, capacity and finance. The first two are combined in what Wood calls the *opportunity frontier* facing the firm, while the third is analysed as a *finance frontier* (Arestis, 1992).

The opportunity frontier depends on the following forces. The demand constraint determines by how much the margin would have to be reduced to achieve a given percentage increase in the rate of sales growth. The more efficient the firm is in its marketing policy, the less will be the trade-off, and the higher the mark-up consistent with each rate of sales growth. But, for any given demand/margin trade-off, we need to consider a third variable, namely, the amount of investment that the firm has undertaken. The greater the investment has been, the lower will be its unit costs relative to its prices, and the greater will be its profit per unit.

The finance frontier relates profits, π, to investment, I. A proportion of profits must be retained for investment purposes. However, not all retained profits are used for investment in real assets such as plant, machinery, buildings and materials. Part of retained profits are used to purchase financial assets, such as shares in other companies, as well as expanding the amount of working capital. The ratio of financial assets acquired out of profits as a proportion of investment, f, means that the total retained profit needed in order to have sufficient funding left over for investment purposes is $(1+f)I$.

Furthermore, not all investment in real assets comes from retained profits. A proportion of investment may be financed by borrowing externally, e. Hence, eI is the proportion of investment externally financed. Hence, internal and external funding is required for any given level of investment, I, so that a retained proportion of profits, $r\pi$, is needed to cover the amount of investment not funded externally. Thus:

$$r\pi = (1 + f - e).I \qquad (9.2)$$

where

> r = the retention ratio
> π = level of profits
> f = the ratio of net acquisition of financial assets to net acquisition of real assets

and

> e = the external finance ratio

Dividing both sides of (9.2) by r, we get the finance frontier, namely:

$$\pi = \frac{(1 + f - e)}{r} . I \qquad (9.3)$$

Equation (9.3) shows the level of profits needed to finance any particular level of investment. In the equation the ratio, f, of financial assets to real investment shows the proportion of liquid assets not currently used by the firm but retained as a form of corporate savings. For the moment let us assume that net acquisition of financial assets is zero, and thus $f = 0$.

Now, let

$$k = I/S.g \qquad (9.4)$$

where

> k = incremental gross capital investment-to-output ratio, (the ICOR).
> S = sales revenue

and $\quad g$ = the rate of growth of sales or demand

Therefore

$$gk = I/S \qquad (9.5)$$

Just as in macro-economic terms the ICOR relates an increase in capital to an increase in the size of the GDP, the incremental gross capital investment to output ratio at the level of the firm relates growth in output (sales) to investment, and hence to finance requirements and profit. Equation 9.4 shows the level of investment required to achieve a unit increase in sales per annum. By definition, $S.g$ is the absolute increase in sales in a period.

Now, dividing both sides of (9.3) by S, and assuming $f = 0$

$$\frac{\pi}{S} = \frac{(1-e)}{r} \cdot \frac{I}{S} \tag{9.6}$$

π/S is the profit margin and if we substitute $I/S = gk$ from (9.5), we get

$$\pi_m = \frac{(1-e)}{r} \cdot gk \tag{9.7}$$

where

π_m = profit margin

Equation 9.7 shows the level of profit margin necessary to provide finance for a particular sales growth rate, g, given the ICOR. It is another form of the equation of the finance constraint.

The opportunity frontier, (9.8), states that the profit margin is a function, u, of the growth in demand and the ratio of incremental gross investment to incremental sales.

$$\pi_m = \mu\,(g, k) \tag{9.8}$$

The opportunity frontier therefore combines the trade-off between demand growth and profit margin with the level of investment expenditure. It specifies the best profit margin that can be achieved for a particular combination of the rate of growth in sales, g, and the investment/output coefficient, k.

In Figure 9.1, the finance frontier is a straight line, sloping upwards as we move away from the origin. The opportunity frontier is a curve, convex to the origin, crossing the finance frontier. Their point of intersection gives us the best combination of the level of profit margin and rate of growth available to the firm. Possible positions are those on or to the left of the finance frontier, and on or within the opportunity frontier.

There are upper limits to the attainable values for the profit mark-up, which are interdependent with the lower limit to the amount of

Figure 9.1   Finance and opportunity frontiers

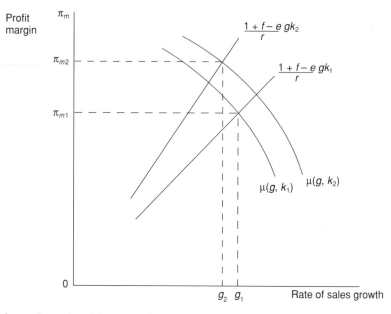

Source: Reproduced from Wood (1975, p. 87).

investment the firm must make to remain sufficiently competitive to survive. The firm will always prefer a strategy which lies on the frontier, to one which lies inside it. Other things being equal, the firm will prefer higher profit mark-ups or sales growth to lower ones, and a lower investment expenditure to a higher one. Positions inside the frontier allow the value of one variable to be improved with no deterioration in the value of the others.

If the value of k increases, the opportunity frontier moves *outward*, because a move to a more capital intensive production method lowers unit cost and thus widens the profit margin at each rate of demand growth. However, the finance frontier moves leftward and becomes steeper, because investment needs per unit of growth increase. The net effect on the rate of sales growth depends upon the balance of strength of these two shifts. The firm, of course, is assumed to be trying to find the highest possible rate of growth – i.e. to maximise g, within the limits of possibility set by the opportunity and finance frontiers.

If, for example, the investment needed for a unit increase in sales is k_1, then the minimum profit margin and the maximum rate of sales growth would occur at π_{m1} and g_1 respectively. If the incremental investment to output ratio then changes to k_2, then both the opportunity frontier and the financial frontier shift in response, and the new profit margin would rise to π_{m2}, achieving a maximum rate of increase in sales of g_2 in this example.

The firm's managers thus have a set of feasible strategies open to them. This is called the *opportunity set* and consists of selecting a bundle of values for all variables, including the price level, output level, marketing level, and investment level. A strategy is feasible at the decision making stage, if the investment involved would provide capacity appropriate to the expected growth in demand. Each strategy is associated with particular values for three *key* variables. The key variables are the profit margin on sales, the sales growth rate, and the level of investment expenditure (Wood, 1975).

The Eichner–Wood model applied to the large construction firm

Values of the investment coefficient, k, are likely to be low in construction compared with other industries. For some firms they may even seem to be negative. A negative value of k would make a nonsense of the idea of a finance constraint, because there would be no need for any finance in order to achieve growth in sales. It is therefore very important to clarify whether k is or could be negative.

The answer lies in the definition of *investment expenditure*. In *investment expenditure* Wood includes all product-development or diversification expenditure, as well as unit cost-reducing expenditure. He therefore includes capital expenditures which raise either *selling efficiency* or direct *cost efficiency*. Put another way, expenditures aimed at improving the firm's competitive advantage are investment expenditures. There is no particular concern with balance sheet conventions as to whether a tangible and disposable asset results from the expenditure. Direct observation of values of investment as defined by Wood is not possible from company accounts. It is, rather, a variable with an estimated value known to the strategic management of the firm – the total amount of strategic expenditure intended to improve the trade-off between the profit margin and sales growth. It is quite possible that this broader investment coefficient could be positive at the same time that an 'increase in the need for (fixed and working) operating capital' was negative.

Suppose k is actually negative for some firms. We then have the result that, with no finance frontier constraint, firms might maximise growth, at the expense of profit margin, up to the point where some other constraint began to exert an influence over decision makers. Such a constraint might be the return on shareholders' funds. A reduction in the return on shareholders' funds, working through the share price, may lead to a threat of takeover by another firm. However, such a constraint is quite weak for construction firms, because of the difficulty predators face in valuing the disposal value of a construction firm's assets, or its profit stream expectations.

In general, for a variety of reasons, contract construction firms have adopted strategies with low levels of investment and low ICORs. This has been mainly in response to uncertainty, and a preference for financial asset accumulation. As a result they have tended to face a poorer trade-off between profit margin and growth than would otherwise have been the case.

Contractors have thus responded to the institutional characteristics of construction markets and the relatively large-number competitive market structure. At around 20 *major* firms, the top end of the UK contract construction market is at the upper limit of numbers for oligopoly.

Rather than a high need for investment funds driving up prices, construction industry firms have tended to adopt strategies for their construction businesses consistent with low levels of investment. Construction industry firms depart from the norm for national industries, and thus constitute a third category of industry in addition to international industries and national oligopoly industries.

Price and output determination for a production-contractor construction firm

For the small firm in construction, decision making, price formation and output determination are far from the description either of perfect competition or of oligopoly given in neoclassical textbooks. Though oligopolistic market structure exists at the 'top end' of the construction market, we consider here also the smaller contractors (though not the smallest, for which see p. 223), who actually undertake production and hire labour.

To see how large and small firms (main contractors and subcontractors) do in fact make price and output decisions, we assume clients ask contractors, and contractors ask subcontractors, to predict demand and cost changes over the life of a project. Contractors then submit *firm price* tenders, without a clause adjusting the contract price to compensate the contractor for all increases in unit prices of resource inputs *ex post* during the life of the project. Tenders requested with such a clause are known as 'fluctuation-of-price' contracts. These assumptions are not necessary to the logic of our model, but serve to simplify exposition and to bring out the impact of uncertain expectations.

Periods

Before we can proceed further, we need to clarify two definitions, both to do with time, the production period and the investment period. Each is an *ex ante* concept, referring to a period over which plans are made by firms. They can therefore also be called the production plan and the investment plan periods. We draw here on Chick (1983).

In each production plan period the firm has a planned volume of output and sales, and a planned mark-up and therefore selling price, and therefore an expected sales revenue. It also starts each period with a known total of fixed costs for that period, and therefore an expected average fixed cost per unit of output; and an expected level of unit direct costs, based upon expected productivity and input-prices. The firm therefore has a predicted gross and net value of operating profit per production plan period. The planned production volume is based upon that of the previous period, adjusted for *ex post* experience from that period (i.e. adjusted upwards if actual demand in that last period exceeded expectation, so that volume could be raised above planned level and operating profit also exceeded expectation; and downward if the reverse). A production period lasts until the firm has sufficient new data and opportunity to review, and if necessary revise, its output, price and hiring decisions.

The investment plan period is that which starts when the firm has the opportunity to review and alter not its output or price but its capacity. This review should be understood to include qualitative shifts in the capabilities and activities of the firm as well as quantitative changes. There will be some time lag before this new capacity can be brought into being, or former capacity disposed of. The capacity of the firm is fixed for the duration of the investment period. In consequence, during an investment period a firm's fixed-cost outgoings are also indeed fixed, at a level depending on those capacity-level-adjusting decisions.

The definition and role of fixed costs

Some fixed costs of a firm are *ex ante*. They are costs that are fixed for the duration of a production period as a result of the implementation of previous investment decisions. Such fixed costs consist of:

- Administration overhead costs, including those costs of ongoing activities only indirectly productive of output, and not therefore proportionate to or directly dependent on the volume of output.
- Debt charges dependent on the pre-existing level of net borrowing; note that because we take a 'cash flow' production period/investment period approach, these count as *ex ante* current production costs because they are actual, unavoidable outgoings in the time period of production, unlike charges on any new debt that may be arranged as a result of forward-looking investment decisions made in the current investment period.
- Lump-sum payments contracted to owners of real assets in use by the firm, such as lease payments on plant and machinery taken on long leases, rent to landowners of leased property, licence fees to owners of technology used under licence.

It is of the nature of fixed costs that they can each be sensibly expressed as a fixed amount per period of time, for a finite series of periods. The number of production periods for which they are fixed is legally a matter of a contract-term (contractual obligations to make payments with unexpired duration of over one production period are fixed costs as far as that period is concerned); organisationally, it also depends on extra-legal matters, so that costs can be treated as fixed for purposes of analysis 'if the firm regards them so'. At the operational level, for example, managerial salary costs of existing staff may be treated by the firm as a fixed cost, even if legally the employment contracts can be terminated within a production period. At the strategic level, which comes into play when investment decisions are made at the start of an investment period, however, these same costs may well be treated as avoidable.

Fixed costs *ex post* are fixed for each production period until the next strategic review or investment period. *Ex post* fixed costs include the consequential fixed costs of the strategic decisions made at the start of an investment period. They include, for example, the debt charges on any new loans incurred to finance that strategy; and the administrative overhead costs of the potentially altered set of activities decided upon.

All actual fixed costs are deducted from the gross profit from current production, rather than treated as part of unit costs. Total fixed costs to be paid during a production period are assumed to be fixed and known at the start of that period. This is a valuable simplification for the analysis, making it possible to disregard such facts as variable interest rate loans.

Certain items are often erroneously lumped in with these actual fixed costs, whereas for the purposes of setting price, output and capacity they are completely different in effect from true fixed costs. The chief such items are depreciation and dividends. Depreciation provisions do not represent actual cash outgoings from the firm. In a cash flow approach, therefore, they disappear. Moreover, neither actual outlays on replacement of worn out fixed capital nor dividend payments to shareholders are contractually fixed or committed in advance, and therefore fail to meet our criterion for fixed cost. The more a firm has financed its investment by borrowing rather than share issue, the higher will be its fixed costs, as defined here. Nor do we regard required return on capital expenditure, out of retained profits, which they may have spent in the past to finance the investment that permits current output, as actual fixed costs of current production. We thus keep a sharp distinction between *profit* and *cost* or *contractual outgoing*.

All this means that we have defined unit costs so as to make them not dependent on the fixed cost per unit volume of output, whilst fixed costs are defined so as to be a known lump-sum per period.

Price and output adjustment between production periods

Gross operating profit (GOP) we define as sales revenue less total direct costs. Indirect fixed costs and taxes are then deducted from gross profits to find actual post tax net operating profit (NOP) per production period. These can be monitored against planned target figures. If there is a shortfall, the only actions assumed possible within the short run are to change selling prices and output volumes. We assume that price elasticity of demand (PED) within an investment period can be either elastic or inelastic according to the circumstances of the firm. Thus gross operating profit will sometimes be increased by raising prices and accepting lower volumes, and sometimes by cutting prices to raise volumes.

The normal post-Keynesian assumption is that a firm can raise its gross operating profit in the short run by raising its prices, but this is limited by the price elasticity of demand. Compared to a high gross margin, a low gross margin requires a much higher price elasticity of demand for the increase in total revenue from a price reduction to offset the increase in total direct cost. A price rise increases the gross operating profit the more the percentage increase in margin per unit exceeds the percentage reduction in quantity sold. In short, a price change affects the gross operating profit depending on both the margin and the price elasticity of demand.

Let

$$m = M/P \qquad (9.9)$$

where

m = gross margin per unit as a proportion of the price per unit
M = gross margin per unit, expressed in the same units as P
P = price per unit

and

$$\Delta P = P_1 - P_0 \qquad (9.10)$$
$$\Delta M = M_1 - M_0 \qquad (9.11)$$

where

ΔP = change in price
ΔM = change in gross margin

so that

$$\Delta M/M_0 = (\Delta P/P_0).(1/m_0) \qquad (9.12)$$

where

$\Delta M/M_0$ = change in gross margin, expressed as a proportion of the former gross margin

and

m_0 = gross margin per unit proportion before price change.

Thus (9.12) shows that for a given percentage increase in P, there is a corresponding percentage increase in M, depending on the gross margin per unit as a proportion of the price per unit.

By definition, and assuming we are considering a price increase,

$$PED = (\Delta Q/Q_0)/(\Delta.P/P_0) \qquad (9.13)$$

From (9.13), a proportional reduction in Q can be shown as

$$\Delta Q/Q_0 = (\Delta P/P_0) . PED \qquad (9.14)$$

Thus for gross operating profit to increase (9.15) must hold true. Namely;

$$\Delta M/M_0 > \Delta Q/Q_0 \qquad (9.15)$$

This is the rule of thumb for the case of a price increase. Naturally, in the case where the firm is considering a price decrease the rule is the reverse – namely, the condition for GOP to increase is

$$\Delta\, Q/Q_0 > \Delta\, M/M_0$$

Table 9.1 summarises the effect on GOP of a given price change.

In Table 9.1, only at a very low margin does the price increase, given the price elasticity of demand, lead to an increase in the gross profit. This is of course most relevant to contractors, because construction firms often work on very low gross margins. It means that it is possible for firms to increase their gross operating profit by a price increase, even when they face an elastic demand for their output, and although a price rise reduces total revenue.

Thus, faced with a shortfall of sales volume and hence GOP below that planned and expected, or faced with an increase in the required level of investment, we might think that construction firms with their very low gross margins would be more likely than others to respond by raising, and less likely to respond by cutting, their prices.

Whether this is so in fact will depend, clearly, on the actual PEDs faced by construction firms. If they have failed to establish customer loyalty, product differentiation and non-price competition, and if they typically face large-number and anonymous competition in tendering,

Table 9.1   The effect of a 10% price increase on gross operating profit at different profit margins, assuming a price elasticity of demand = –3

Profit margin at start	$M_0 = £40$		$M_0 = £25$		$M_0 = £5$	
	t_0	t_1	t_0	t_1	t_0	t_1
Price	£100	£110	£100	£110	£100	£110
Quantity	1,000	700	1,000	700	1,000	700
Total revenue	£100,000	£77000	£100,000	£77000	£100,000	£77000
Cost per unit	£60	£60	£75	£75	£95	£95
Total cost	£60,000	£42,000	£75,000	£52,500	£95,000	£66,500
Margin per unit	£40	£50	£25	£35	£5	£15
m_0	0.4		0.25		0.05	
$1/m_0$	2.5		4		20	
GOP	£40,000	£35,000	£25,000	£24,500	£5,000	£10,500
ΔGOP		–£5,000		–£500		£5,500
$\Delta M/M_0$		25%		40%		200%
$\Delta Q/Q_0$		30%		30%		30%

then their PED will actually be very high and therefore elastic. If, on the other hand, they win much repeat custom by negotiation, compete on non-price criteria, such as quality, in any of its relevant meanings, or speed, and face small-number competition from a set of known close rivals, then their PED may be much lower and therefore relatively inelastic.

The price and output decision within a production period is essentially a selection of the gross mark-up, in the light of the expected impact of price on output volume and on net operating profit. Figure 9.2 illustrates the position of a firm given a choice of either a lower price at E or a higher price at A. The gross operating profits are shown by the heavy horizontal and vertical rectangles. At price E, the firm sells P units. Total revenue is represented by EGP0. Total direct costs are represented by KMP0. The margin is KE. The gross operating profit is EGMK. Total indirect costs are HJMK. Net profit is therefore EGJH. Now, assume the firm selects the higher price at A, and that sales fall from P to N. Total direct costs are reduced to KLN0. The new margin is KA, and the new gross operating profit is ABLK. The indirect costs spread over 0P in rectangle HJMK, now have to be spread over 0N. Therefore the indirect costs of IJML are moved to CDIH. The new net profit is therefore ABDC.

Figure 9.2　The effect of 2 prices on gross and net operating profits

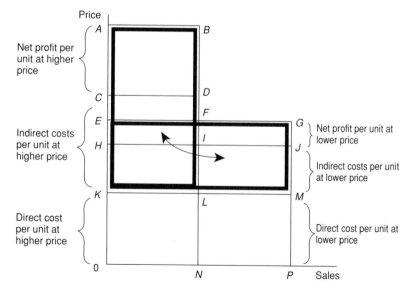

This decision for each period is linked in a chain not only with past output and investment levels but also with future ones. The past decisions will have affected the firm's existing commitment to fixed costs, and therefore the GOP required merely to cover these even before there is any net operating profit (NOP).

Figure 9.3 illustrates the relationship between: output, Q; GOP; total revenue, TR; and total direct cost, TDC. At output Q_1, total revenue from sales is A, total direct cost is B, and total gross operating profit is AB (equal to $TR_1 - TDC_1$), the distance between the TR curve and the TDC line, at output Q_1. Total cost, direct and indirect, is shown by the line TC. At output Q_1, total cost is CQ_1 (or TC_1), equal to TDC_1 plus TIC, a constant. Net profit is AC, equal to $TR - TC_1$.

In Figure 9.3 the gross operating profit is the vertical difference between the TR curve and TDC. To begin with the GOP increases as output increases and then declines. At low levels of output the GOP may even be insufficient to cover indirect costs (TC > TR). At high levels of output the PED becomes inelastic when price is reduced. As a result TR declines if firms attempt pricing strategies to increase sales in the short run beyond a certain level.

The financing and demand constraints on growth have to be traded off against one another, if the firm is seeking to raise internal finance out of an increase in its GOP by increasing its prices and thus reducing its sales volume. On the other hand, if GOP can be increased by cutting prices, there is no contradiction or trade-off. The same actions

Figure 9.3 Gross and net operating profit, total revenue and total direct and indirect costs, for a production period.

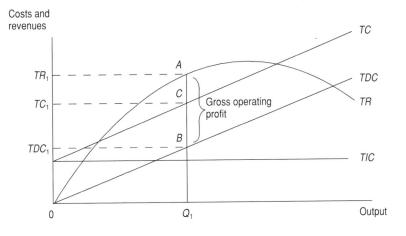

both raise the finance necessary to allow growth and provide the necessary increase in demand volume.

Firms enter a given production period with a certain gross mark-up for each business activity, and therefore a certain set of prices. This implies that with these mark-ups they expect to achieve a certain value of turnover, and therefore a certain gross operating profit. If these expectations are fulfilled, the firms achieve their predicted NOP, because fixed costs for the period are known in advance.

The price and output adjustment decision of specialist subcontractors

It is possible to apply this model of pricing, percentage margin and the PED to a specialist subcontractor. Let us suppose a *unit* of output refers to the smallest quantity of work that a specialist subcontractor would undertake as a separate subcontract, so that the smallest potential marginal unit of increment in volume is 1 unit. We might further imagine, however, that its typical job comprises 20 units, and that it planned to carry out 50 jobs in the production period, a total expected output equivalent to 1,000 units. In the terms of our example above, suppose the firm thought its PED to be at least 3, and set a 33 per cent mark-up on direct costs of £15 per unit, selling its output at a price of £20 per unit.

Now, if the actual sales volume of this firm proves to be less than expected, the shortfall will bear heavily on its net operating profit. Say the firm only sold 900 units of output at £20 per unit, instead of the expected 1,000 units. Its actual total revenue would be £18,000 instead of £20,000. Its total direct costs would also be proportionately less than expected, at £13,500 instead of £15,000.

As a result, Table 9.2 shows its gross operating profit is therefore £500 below target, at £4,500. Now, suppose total fixed costs for the production period are £3,000. The net operating profit is therefore reduced from £2,000 to £1,500. A 10 per cent shortfall in sales has reduced net operating profit by 25 per cent. If the tax rate is 30 per cent, the post tax net operating profit would be;

£1,500 (1 − 30/100) = £1,500 − £450 = £1,050

instead of the planned figure of;

£2,000 (1 − 30/100) = £2,000 − £600 = £1,400.

Now let us suppose that the firm had planned to distribute £700 and invest £700 of its expected post tax net operating profit in increased

Table 9.2   Effect of lower sales volume on net operating profit

	Planned	Actual
Sales volume	1,000	900
× Price	£20	£20
= Total revenue	£20,000	£18,000
− Total direct costs	£15,000	£13,500
= Gross operating profit	£5,000	£4,500
− Total fixed costs	£3,000	£3,000
= Net operating profit	£2,000	£1,500

working capital. How might the firm react? It has two decisions to make. The first concerns the use of its net profits. The second decision is about its prices. Much will depend on the firm's diagnosis of the reason for the shortfall in its volume of sales.

The firm might conclude that there had been an overall drop in demand in its market. Evidence for this might include hearsay that rivals had experienced similar shortfalls, a reduction in the number of invitations to tender, but no reduction in tender success rate, and published indicators or forecasts of demand, including orders and trends for construction. If the firm decides to respond by leaving its prices unchanged, it will then expect a similar or even lower volume of sales in the next production period. It may decide that it no longer needs as much operating working capital as it planned. It could then retain its planned dividend at £700, and invest only £350. Even this £350 might be used to acquire financial assets. Alternatively, the firm may also consider the option of responding to the diagnosed drop in demand by cutting its prices, so as to return towards an output volume of 1000 units.

As stated earlier, the firm believes its PED is around 3. For every 1 per cent *increase* in the firm's prices relative to those of its close rivals, there would be an expected drop of at least 3 per cent in its sales volume. However, the PED of a *reduction* in price may be quite different. This is analogous to the kinked demand curve of oligopoly theory. However, the reasons for differences in PED, depending on the direction of price movements, are not confined to the assumed imitation or non-imitation by rivals. We go further, in assuming that firms have only very sketchy ideas about PED, so that they are in no position to maximise profit. Moreover, we make the assumption that firms have a strong bias against experimenting with their prices, and leave them alone so long as expectations are fulfilled. However, for the moment we will assume PED is the same regardless of the direction of price changes.

If the firm, therefore, thinks that its rivals will not respond to the fall in demand by changing their prices, this is the relevant PED – provided it believes that its demand curve has shifted but not changed its slope. Assuming price changes will only be considered in amounts of at least £1 per unit, a price reduction will only raise GOP if volume increases to more than 1,125 (the amount which, multiplied by the new margin per unit, £4, gives, not the old forecast GOP of £5,000 but the new *ex post* GOP of £4,500.) This requires an increase of 225 units from last period sales of 900 units, i.e. a 25 per cent volume increase in response to a 5 per cent reduction in price – a PED = 5.0. Thus this response would be rejected.

Perhaps, instead, the firm might try to improve its situation in the new conditions of lower demand by raising its margin and price? If it raised its margin per unit to £6 at a price of £21, in order to achieve a GOP of at least £4,500 it would need to sell at least 750 units. The GOP is found by multiplying the volume of sales by the margin. Thus quantity must fall by no more than 150 units from a base of 900 (i.e. 16.7 per cent), in response to a price change of 5 per cent – a PED of no more than 3.33. This is quite close to the 'guesstimate' of PED as 3. If actual PED is at the lower end of its estimated range of values, therefore, this policy might work.

Thus, apparently perversely, a response to reduced demand of increasing prices is more plausible as a way of restoring GOP than one of reducing prices. But, the most likely response is that the firm will leave its margins and prices unchanged, and absorb the reduction in GOP in lower retentions and investment.

Using the same methodology, it is possible to derive strategies in response to a variety of circumstances under different trading conditions and for both elastic and inelastic demand. For instance, a situation may arise where the firm's diagnosis is that sales volume has fallen below target because of the unpredicted behaviour of rivals. Another scenario may occur where a firm has lost sales because somehow it has allowed its unit costs to drift above those of its rivals.

The price adjustment decision of a major main contractor

One of the special characteristics of the construction industry is the divergence of cash flow from profit in significant ways, especially for the major contractors. Ive (1994) and Punwani (1997) have shown that, for the largest and most powerful main and management contractors, their business has the peculiarity that need for operating capital changes inversely with turnover and output volume. That is, when their volume of output is growing through time, in each period they need less operating capital than in the one before. Indeed their operating capital can be actually negative – a result of operating with negligible fixed assets and negative working capital. The latter is negative because cash flow in each production period is highly positive – receipts greatly exceed outgoings. Thus, far from having to inject money capital in order to operate the business, the activity finances itself, and more.

Need for working capital is negative because the firm is collecting money from its construction clients faster than it is paying it out to its trade contractors. At any one moment the firm's balance sheet therefore shows an overhang of liabilities owed to creditors in excess of the firm's current operating assets. Ultimately, if the firm were to cease trading, it would therefore undergo a period in which its outgoings (settlement of outstanding debt arising from past operations) would greatly exceed its income (collection of monies due arising from past operations).

The same, to a lesser extent, would be true if its output shrank in scale. As long as it continues to expand, however, cash-flow will be both positive and increasingly large. Operating profit margins, gross and net, may be quite low as a proportion of turnover or as a percentage mark-up on direct cost. However, because the cash surplus can be used to acquire financial assets, and these in turn can yield non-operating income, non-operating profit from interest, dividend, and rent income can be large.

Now, such a business clearly faces no financial constraint, in the usual sense, on its growth. Moreover, the demand constraint may well be dealt with by reduction of margin, up to the point at which this ceases to raise combined profit (operating and non-operating). This would suggest a tendency towards lower margins. There is evidence that in the last two decades, the largest contractors have been cash-generators. At the same time, they achieved rapid growth relative to the total market, increasing their market share, having cut their net operating profit margins (Stumpf, 1995; Punwani, 1997).

The last point needs elucidation. Company accounts data do not always, and business ratio data hardly ever, report the gross margin (gross markup on direct cost). Instead, they report the *ex post* net margin, expressed as net profit over turnover, and it is this margin which we can observe to have been falling, or at best constant, for the largest contractors even during the strong boom conditions of the mid- and late-1980s. This certainly suggests that their gross margins were also falling, for it is unlikely that fixed costs were falling as a proportion of turnover at that time – considering that large firms were increasing their administrative, professional, technical and clerical staffing faster than their output volume, and that salary levels of these office employees were rising (Ive and Gruneberg, 2000). For big main contractors, devoid of debt and with low depreciation costs, staff costs are undoubtedly the main component of fixed cost. The trend movement of contractors' fixed costs is likely to be determined by the trend in their staff costs. The data problem for measuring gross margins lies in lack of consistency and detail in firms' reporting of costs, and especially their division into direct and fixed components. Many firms report staff costs, in particular, together with direct costs in a single reported figure for cost of sales.

Now, let us consider the response of such a firm when demand falls below the level expected. Large main contractors' sales divide in most cases into a major (though possibly diminishing) part in which volume is highly sensitive to relative price, and a minor (though possibly growing) part which is less price sensitive, because these sales are competed for on grounds of unique technical competence or distinctive competitive advantage in quality, or as a design-and-construction or total finance-design-and-construct package. If the firm leaves its prices unchanged in response to the fall in demand it will certainly face sales volume below that planned. The extent of the shortfall in demand depends on whether close rivals cut their prices.

Now, so long as the real volume of sales is still increasing, price undercutting by competitors may be tolerated. But, if the real volume of sales starts to fall, with rival firms cutting their prices, then the firm will be faced with a major cash-flow problem. Punwani (1997) shows this effect in operation for the largest contractors during the downturn of 1989–91. It is then likely that the firm will respond by cutting gross mark-ups, sharply if necessary, in order to maintain at least a steady real volume of sales and output, and hence prevent a sudden need for an injection of working capital. The alternative for the firm is to find the finance for such an injection either by disposal of assets in other businesses or by issuing new share capital, typically via a rights issue to existing shareholders. Selling assets in other businesses is an unattractive option if those other businesses are simultaneously facing a disappointing level of demand. Their working capital will be tied up in illiquid stocks only sellable if at all at deep price discounts. Issuing new share capital may likewise involve a heavy cost if the firm's present share price is depressed.

The pricing and output behaviour of a near-firm or micro-firm

We have dealt above with the capital investment, financing, pricing and output decisions of both the larger and medium-sized construction firms, which we assume to be growth orientated. A different approach is needed to understand capital investment by the smallest *near-firms* or *micro-firms* (see Chapter 3, p. 81, for definitions), whose owners see profit mainly as distributable income for themselves, and whose business objective is often just survival.

For near-firms, as with the large firms, the maximisation of the return on capital employed is a mistaken basis for a model of their behaviour, but for a different reason. This time, it is because the owner producers are trying to *maximise* their total short run income. This income is a mass not a rate, and includes income other than profit, such as salary and owners' benefits in cash and kind as well as income derived from profits. They are reasonably described as short-run *maximisers*, subject to the caveat of bounded rationality, because they see the long run future as so hazardous and outside their ability to control. In any case, they see no effective way of securely increasing their future expected income by systematic re-investment.

It is because they do not accumulate capital that they are only near-firms and near-capitalists. They do from time to time make gross fixed investments, but they do not plan for them. They do so only when forced in order to retain an operational capability, mainly replacing fixed assets that have actually worn out. They often only carry out these investment decisions long after the assets may have become economically obsolete, in the sense of being less profitable to use than possible replacements.

Consider a schematic example of such a micro-firm. We have placed our exemplary firm at the larger end of the range covered by this type of entity, so that all conclusions drawn will apply *a fortiori* to smaller near-firms.

Assume the firm has a small board of directors, comprising its two founders plus members of their families, who may also draw fee income from the business as part-time directors. It is a private limited company with a nominal issued capital of £20,000. This capital, together with a bank loan of £20,000, secured on the houses of the owners, is used to own a second-hand van, a personal computer, and some items of construction equipment. The greater part of the £420,000, however, is used to cover advances of wages and payments to materials suppliers.

Members of the board comprise the entire management of the firm, and as such draw salaries. Board members do not work as direct producers. The business is run from the homes of the owners. Clearly, when first set up, any *profit on shareholders' funds* is likely in absolute terms to be negligible in comparison to managers' salaries, however high the relative rate of profit on capital employed.

For example, a rate of profit of 50 per cent would be a flow of £10,000 per annum, whereas we may assume the directors draw combined salaries of £60,000, plus directors' fees of a further £10,000. Indeed the directors have an interest in minimising reported profit, since distributed profit will be doubly taxed compared with *salary* income. Insofar as the operations of the firm yield an economic surplus to the owners, they will, for tax reasons, take or appropriate that surplus mostly in the form of exaggerated salaries – that is, salaries above the level they could command in the labour market as employees. Let us suppose that declared profit is as high as £10,000 p.a., that the salary exaggeration is £20,000, and with directors' fees of £10,000, the total annual surplus is £40,000.

On average at any one time the firm has, say, ten direct producers working for it. Annual gross payments to these workers total £150,000, the equivalent of an average annual wage cost per worker of £15,000 gross. The business operates mostly as a subcontractor but also, sometimes, as a *jobbing builder* appointed directly by small private clients.

Its gross value added should therefore be computed as £150,000 (operatives' wages) + £40,000 (manager's salaries) + £40,000 (owners' surplus) = £230,000 p.a. Its rate of surplus is somewhere between $(40/190).100$ per cent and $(80/150).100$ per cent, depending on how we decide to think in this case about the managers and their salaries – as part of the workforce or as agents of 'capital' (see, e.g. Marglin, 1974). Let us take a middle course, and treat only the exaggerated part of management salaries as part of 'capitalist income', but exclude managers' salaries from the denominator; the rate of surplus is then $(40/150).100$ per cent = 27 per cent; and the amount of surplus per worker is £4,000. Its annual turnover is, let us say, twice its gross value added, or £460,000; and therefore its annual production-cost outgoings other than on wages are £230,000 – mostly comprised of payments for construction materials, but also including office-consumables and hire charges on items of equipment.

From the declared gross profit of £10,000, the firm must pay a fixed interest cost of £2,000, assuming interest is charged at 10 per cent. Net taxable profit is therefore £8,000. Let us suppose that tax is paid on declared net profits at 25 per cent. Post-tax net profit is then £6,000. All of this, £6,000, is then distributed as dividend, giving a dividend yield of 30 per cent on shareholders' funds.

In any one week the average value of outgoings is £380,000/52 = £7,308, of which the weekly wage bill is £150,000/52 = £2,885, and the weekly bill for materials, etc. is £230,000/52 = £4,423. Assume that main contractors and clients pay the firm on average six weeks after the work is done, whilst wages are paid one week in arrears, and materials bought on two weeks credit and 'just in time'. At any one time the near-firm would need to be able to finance four weeks' worth of use of materials, and five weeks' worth of labour costs. This would amount to (5 . £2,885) + (4 . £4,423) = £32,117. This need for operating working capital will be the main single use of the firm's money capital and overdraft, together totalling £40,000.

In addition the near-firm will need a cash reserve of several thousand pounds to enable it to continue to pay wages and materials' suppliers in the event of late or non-payment by a client, and to meet any unforeseen bills. The firm's cash reserves may be held in its current account balance at the bank, or as in this case, may be the unused remainder of its overdraft limit. The rest has been used as fixed capital, to purchase the van, the computer, etc.

The firm's pricing decisions are detached from long-period investment planning, simply because such near-firms do not engage in the latter. Instead, they are constrained by a chronic shortage of operating financial capital.

Faced with demand conditions that permit expansion, the near-firm may reject the option because it lacks the necessary working capital, or the means to obtain more. If it accepts the option, it may well go bankrupt through *overtrading*. Overtrading occurs when a firm has a level of turnover and, therefore, levels of debtors and creditors too high relative to its capital base, even if the extra output would eventually have been highly profitable, when all monies had finally been paid and received.

Faced, on the other hand, with a reduction in demand, the near-firm has a range of feasible responses it can make. It can decide to cut its volume proportionately or more than proportionately. Or, it can decide to cut its prices in an attempt to sustain volume. If it reduces its revenue, this may shrink to the point where the near-firm can no longer sustain the annual outflow of inflated salary to its partners or directors. At any rate, this fixed appropriation on the gross trading surplus will rise as a proportion.

It is reasonable, in fact, to argue that, for the near-firm of this type, its *survival and distributed income* aims translate into an objective of maintaining sales volume as stable as possible. However, because it lacks substantial balance sheet reserves, it cannot allow itself to make losses in its profit and loss account. Thus, in the end, owner distributed income may have to be reduced. In a severe recession this may result in the partners or directors choosing to cease trading, if income from the near-firm drops below the actual opportunity cost of their labour.

Concluding remarks

The Eichner–Wood models sets out clearly how firms can plan their investment, sales and pricing decisions in a consistent model of the firm.

Profits depend on firms' gross profit margins, as does (but inversely) their rate of growth in sales. The profit margin consistent with any given rate of sales growth will be higher, the greater the firm's past investment in competitive advantage.

But, the greater the rate of sales growth and the higher the firm's investment-to-output ratio, the larger will be the amount of finance required to fund expansion. We show that there is a finance constraint on growth, given by the need to finance a constant proportion of investment from retained profits, and ultimately limited by the present level of profits.

In formulating their pricing strategies, like firms in most industries, firms in construction first calculate their unit direct costs. A mark-up is then added to these direct costs to give a selling price. The volume of sales multiplied by the mark-up produces the gross operating profit. Due to the tendering process mark-ups are often driven down by the need to be competitive, but when gross profit margins are already low the price elasticity of demand needed, to win sufficient extra sales so that gross operating profit does not fall despite the reduction in margin, needs to be correspondingly high. Otherwise, the reduction in price and increase in sales will generate insufficient additional revenues to meet the increased costs of increased output. Strategically, firms must struggle to reach, and then to stay on, the path of high investment/competitive advantage/widening margins, and balance of need for and supply of finance at a high and steadily rising level.

Most construction firms fail to achieve this, and are instead driven onto the alternative path of low investment/no competitive advantage/narrowing margins, and balance their need for and supply of finance at a low level.

10
Marketing and Production Decisions of Speculative Builders and Contractors

Introduction

This chapter discusses the marketing, output and pricing decisions of speculative builders and contractors. We look at the marketing and production considerations which speculative builders need to take into account in planning their production. These include the relationship between their fixed cost purchases of land and their variable costs of building using specialist contractors. We also look at the marketing and production considerations of contractors, who are involved in tendering procedures, and note how contracts are let sequentially throughout the year. This enables contractors to adjust their tender prices in response to uncertainty and evolving conditions.

Speculative builders

The price and output adjustment decision of speculative builders

The investment appraisal for a private sector project is always a question of the amount by which projected benefits are estimated to exceed projected costs. Costs have a similar meaning for a building user or a speculative developer. For the owner or user, however, the idea of *benefits* refers essentially to the financial benefits that flow from the use of the building, such as greater output, more efficient production, etc. For the speculative developer, by contrast, the yardstick of benefit is the selling price or rental value of the building. Only direct monetary costs or benefits to the decision maker are taken into account. The building represents a sum of money, an investment of capital, and is regarded in terms of its money profitability to its owner.

A simple way to conceptualise the investment decision is in terms of an expected profitability threshold, such that all investments offering an expected rate of profit in excess of this level are undertaken, whilst those that do not are not. This threshold will rise and fall over time so as to balance the demand for and supply of investment funds.

For the speculative builder, the expected rate of profit on a development may be derived as follows:

$$\pi_e = P_e - C_e \tag{10.1}$$

where

π_e = the expected mass of profit
P_e = the anticipated selling price of a building
C_e = estimated cost of production

From the anticipated selling price of a building, deduct the estimated cost of production, including the purchase cost of land. This gives the expected mass of profit. Expressed as a ratio of anticipated capital employed this gives the projected rate of profit as shown in (10.2):

$$r_e = \pi_e/K_e \tag{10.2}$$

where

r_e = the projected rate of profit
and
K_e = anticipated capital employed

If this exceeds the *target* or *threshold* rate, the developer may be expected to go ahead.

The volume of output activity initiated by speculative builders in total, therefore, depends upon the number and size of investment projects that each developer identifies as likely to yield in excess of this target rate. This is broadly analogous with the orthodox idea in which the marginal efficiency of capital or the marginal return on investment is brought to equality with the cost of capital to the developer. The cost of capital to the highly geared – i.e. mainly debt-financed – developer is determined by the rate of interest, after making what is called a *risk adjustment*.

The cost of capital to the developer is supposed to incorporate a premium reflecting lenders' accurate assessment of the extra risk of non-repayment involved in lending to the developer, rather than purchasing a default-risk-free financial asset, such as Treasury bonds. This *risk* is supposedly calculable in terms of probability. It derives from two sources, namely the project itself and the financial nature of the borrower. The project risk is the risk that the particular development project in question will in the out-turn yield losses rather than profits. The financial risk is concerned with the risk that the borrower is, in the sum of their activities, so highly geared and illiquid that they will be unable to *cover* a project's loss with net income from other activities or the proceeds from sales of other assets.

The total level of demand for funds from speculative developers may, in some circumstances, be sufficiently large to affect the base rate of interest. However, it is rare for such a mechanism to be sufficient to choke-off a boom in speculative development.

Where the developer is also the builder, i.e. managing their own construction projects, no market prices exist for construction projects as such, for there are no transactions in contracts to build whole projects. Instead, there are the prices charged by specialist contractors and other suppliers. Equilibrium construction prices and property market prices would be those which yielded developers the average rate of profit on new investment so that there was no net flow of capital into or out of development activity. Total investment activity by developers would follow a steady state path. This is the condition of long-run equilibrium. However, the speculative development market is never in long-run equilibrium. Instead prices are either rising or falling. In the short run the level of new output of projects produced by developers may be too small in relation to the total built stock to eradicate an imbalance between the demand and supply. In the longer run, on the other hand, there may quite possibly be overbuilding in a 'boom', so that by its end supply overshoots demand for space.

What is most important from a construction market perspective is that there is no real sense in which the level of selling prices for the speculative built stock is determined in the long run by the level of construction costs plus the average profit mark-up. Not only does the price of land intervene to break any close relationship between construction cost and building price, but also the level of new output is itself not very sensitive to a change in the level of construction costs and prices. Rather, new output is more a function of expected building (property) prices and expected interest rates as well as exogenously determined public and private sector building requirements. Thus the supposed equilibrating mechanism, by which changes in the gap between the prices of speculative buildings and construction costs would be self-righting, does not function.

In the short-run, however, construction prices fluctuate to whatever level is compatible with the level of output desired at that moment by developers. Meanwhile, prices of newly built stock, like those of the rest of the built stock, depend in part upon the demand and price expectations of speculative holders of that stock, and in part upon non-speculative building-users' estimates of their own future need to occupy built space relative to their present occupancy, given assumptions about income and relative prices.

The balance of speculative property market sentiment lies between property market *bulls*, those who think prices of property will rise over time and therefore wish to add to their property holdings, and *bears*, those who think prices will fall and therefore wish to reduce their holdings. These demand-side expectations and user needs can swing and change with a much greater magnitude than can the supply of built space. Whilst property prices are therefore affected by both supply and expected demand, it is clearly the latter that predominates in the sense of *turning* the direction of change of prices. Where there are tight restrictions on new output, as for instance land use planning regime restrictions, property indeed becomes a quasi-scarce good.

The output of an individual speculative house building firm, in the short-run, is not really limited or determined by the eventual tendency of unit costs to rise as output rises in the short run, because some inputs are in fixed supply, causing diminishing marginal returns. Marginal costs, in the sense of variable production costs per dwelling, certainly do rise eventually as each firm attempts to expand its output.

However, it is simply not the case that the marginal cost of constructing a dwelling rises to equal selling price or marginal revenue per dwelling. Rather, what may, after a substantial lag, restrict the output of smaller speculative builders as a 'boom' nears its end is their inability to finance the purchase of sufficient replacement land to permit them to reach their target or desired level of output, especially as these small speculative builders do not tend to possess multi-year land banks. Margins may or may not at some point get squeezed between acceleration in the rise of construction costs and deceleration in the rise of house prices. In any event, as land prices soar after several periods of expansion, the gross surplus of house price over construction cost will fail to widen equally fast. Consequently, retained profits from the sale of current output become insufficient to finance the further purchase of land on which to build planned output. This coincides, with eventual unwillingness of lenders to increase still further their exposure to possible default by speculative developer borrowers. Thus the combined total of internal and external finance can only be increased at a rate which imposes an effective finance constraint on the ability to expand output.

It would seem that it is this financing constraint, rather than any ability to foresee downturns in the housing market, that saves at least some housebuilders from over-extending themselves in terms of output to the point at which demand downturn would mean bankruptcy.

All the above would apply in its broad outlines to the situation where selling prices of the commodity (the building) are taken as given to the firm. The firm then only has to decide on its volume of output. This situation is called price-taking by the firm. It is relatively rare, and can apply only to homogeneous or 'graded' commodities, such as speculatively built housing, and even then only by dealing in averages, and ignoring the spread of actual house prices around these means.

Note that, in price taking markets, firms do not set their own profit margin. On the contrary, this is determined retrospectively, after it is known what actual selling prices are. Unlike firms in price-setting markets, therefore, producers will only produce if they perceive that there is normally a wide gap between their costs and the selling price, to minimise the danger of loss.

For each firm, the amount of profit is relatively uncertain and prone to wide short-run fluctuations. However, the only sources of difference in gross profitability between firms lie in their having different levels of cost of production or, and this is much more important, in the different levels of success with which they time their purchases and sales. In this again, it differs from the price-setting type of market.

In the price-taking kind of market just as much as in price-setting markets uncertainty is prevalent, though it takes different forms. Output decisions by the firm are made on the basis of expectations about future levels of price and cost. These decisions are of two sorts. Short run decisions concern the volume of immediate supply, and long run investment decisions concern future capacity. Both concern the volume of output. Out-turn profitability depends upon the levels of prices and costs that eventually prevail when output has been produced. In price-setting markets, in contrast, uncertainty takes the form of unpredictability in the *ex post* volume of output or sales, rather than uncertainty of prices.

For the price-taking firm nothing is accurately known in advance. A price-taking developer cannot know its selling prices, nor the level of average cost it will incur for its production, since this depends on how market conditions evolve. However, price-taking implies that the firm's output is insignificant in relation to the total supply, so that its output decision has no feed back effect on the market price. The overall degree of uncertainty is however high. Profit uncertainty is greater, the more unpredictable the market price, the more unstable the production unit costs, and the longer the time lag from decision to produce until time of sale.

For a speculative builder, ignoring at this point the replacement cost of land, total gross profit is total revenue minus total construction cost. Total revenue is market price multiplied by volume, while total cost is the average construction cost multiplied by volume.

Application of the price-taking average unit direct cost model to a speculative house builder

House builders are multi-product, multi-establishment firms. This is so at the strategic level, where the firm produces a product range consisting of lines targeted to market segments. At the tactical level the firm possesses a set of sites each capable of yielding a variable annual or monthly flow of new dwellings, and each site represents a production facility or establishment.

The sites of a speculative builder are, in cost terms, rather like the projects of a contractor. There is a certain site or project overhead cost, the annual value of which is not proportionate to or dependent on output volume on that site or project. This overhead cost, such as the land price of a site, can be avoided only by not commencing the project. Because the site or project overhead cost is not dependent on output volume it is in a sense a fixed cost, but because the cost can be avoided they are not like head office or central overhead costs. However, the difference between the contractor and the speculative builder lies in the duration of the period for which these project overhead costs are fixed.

Let us divide a speculative construction firm's investment into strategic and project investment. The former comprises expenditures not related to specific projects, and undertaken to increase the general capacity or competitive advantage, either by raising marketability or by reducing the general unit operating costs, of the firm. The latter comprise expenditures made necessary by the decisions to undertake particular new projects.

In the case of a house builder, the most important of these project investment expenditures will be land acquisition expenditure. (This is not necessarily the largest item of project-investment, but it is the longest lived item, and this, together with its size, makes it usually the most important.) Thereafter, there are expenses of securing and maintaining the site, which are fixed costs with respect to project output volume per period. These costs follow from the investment decision (decision to acquire the site).

Then there are the costs of setting up production management for the project, once production starts. Production management costs have both a time-related and an output-related component, and are therefore a mixture of fixed and variable cost, as far as their relation to project output volume per period is concerned. Next there are direct production costs, chiefly payments to subcontractors and (if labour-only subcontractors are used) to suppliers of construction materials. These costs follow from the decision to produce a certain volume of output in a time period.

Any decision to produce implies a need to inject working capital into the project. This may be called the project need for operating working capital. Now, we will assume that when project investment decisions are made there exist project production and sales plans. The production plan specifies volumes of output from that project in each of a specified series of time periods. A project sales plan shows the intention to generate specified sales revenues in each of a series of time periods. Sales and production plans are linked by assumptions about time lags between production and sale of output. From these, together with further assumptions about trade credit periods available from suppliers, estimates of the project need for operating working capital for each period can be derived. This estimate of the project need for operating working capital is taken into account in making the project investment decision. A firm would wish to ensure it could finance the total capital consequences of a project before committing itself. However, the firm would not necessarily make an actual investment of most of this working capital into a project at this early investment decision stage.

At any one time, the firm's operations will be dominated by an inherited set of projects, resulting from past investment decisions. Therefore, the total need for operating working capital in the medium term can only be controlled by adjusting production plans on existing projects. In this the speculative builder differs from the contractor, whose inherited portfolio of projects, in any quarter-year, say, may mostly be planned to finish within the next few quarters, with new projects normally starting in each quarter. In contrast, the evidence for larger speculative house builders is that a typical site will be built out only over several years – sometimes as long as ten years – and that site acquisition is very lumpy relative to output flows.

Thus it is important to the house building firm to keep its production costs per period, on each project, variable. This means not entering into long-term, fixed volume supply contracts – since it is such contracts that turn a potentially variable cost into a fixed cost. Short term spot contracts may, however, be inefficient in terms of transaction costs. Nevertheless. there will be a tendency to prefer such inefficiency to cost fixity, as the lesser of two evils, especially as the size and source of transaction costs may be disguised by the accounting systems used.

The relative size of the firm's central overhead, as opposed to project related overhead, will depend both on how many strategic investments it undertakes, and upon its practices with regard to centralisation of certain management functions, such as purchasing of materials. Pecuniary economies of scale from central purchasing, through maximising bargaining power, and real economies of scale, for example through minimising waste or minimising capital tied up in stock holding, may be sufficient to offset the general desire to minimise fixed management costs. However, because of the typical long duration of projects, even project overheads must be regarded as fixed costs in all but the long period.

Speculative house builders tend to have rather large total fixed overhead costs per unit, and also rather high normal gross mark-ups on unit direct cost. Their unit direct cost schedules tend to be flat with respect to their own volume until a full capacity volume is reached. This flat unit direct cost nevertheless shifts in response to aggregate market demand conditions.

Shape of the demand schedule facing the speculative builder

The aggregate or industry demand curve for private new housing is clearly downward sloping. The quantity demanded responds to price, especially if price is measured as the price of new housing *relative* to an index measuring both the price of existing dwellings and the cost of housing refurbishment. However, the assumption above is that each house building firm faces a horizontal demand curve, which is, of course, the standard assumption of perfect competition. That is, the assumption is that the firm will not have to reduce its prices in order to sell a larger volume of its own output. Such a situation is conceivable, if the output of a firm (Q_f) is minute relative to the total supply of all houses for sale on the market (ΣQ), and there is no product differentiation, plus a frictionless marketing process.

Now, it is probably true to characterise the market for new housing as containing a very low degree of firm-specific product differentiation. However, it is doubtful that for most firms Q_f is minute relative to ΣQ. For a firm such as Wimpey, this is clear enough. But even smaller builders may produce significant shares of total supply in their own local market areas.

Is the firm's demand volume price elastic or inelastic? We incline to the view that, in most cases, the speculative house builder faces a rather price-elastic but kinked demand curve, so that the volume of sales per period will vary quite sharply in response to an upward change in the firm's house prices relative to prices of other supply on the market, whereas a cut in relative price may yield less response. The question, however, is an under-researched one.

Even more to the point than significant market shares, is the fact that new housing is sold in a real market institutionally and informationally very different from that of the Walrasian fable, as discussed in Chapter 2 in terms of the social structures of speculative housing provision. Thus, our reasons for making the contrary assumption, of downward-sloping demand curves facing the speculative building firm, should by now be clear. This would require modification of the model set out above, along the lines suggested in Chapter 2.

Contractors

Shape of the demand schedule facing the contracting firm

What does it mean, to say that, at the prevailing market price, the small contractor can sell any quantity they wish, without an upper limit, so that their output decision is simply determined by the point of intersection of their marginal cost curve and the price line?

Invitations to tender

In reality, in a construction world dominated by one or other form of selective competitive tendering and negotiation, no contractor can undertake a larger volume of work than that for which they receive invitations to tender. They depend upon clients to short-list them as tenderers for a project. If open competitive tendering were the norm, then in principle each firm could tender for as much work as it liked, up to a maximum of the total volume of projects put out to tender by all clients in that time period. Contractors expend marketing effort trying to increase the number and value of invitations they hope to receive in future, but this effort takes time to work. Thus, in the short run, the value of invitations is a given, outside their control. For each price level, we can convert this value into a volume.

Graphically, we can represent this by drawing a vertical section of the firm's demand curve. This is the maximum amount the firm could sell even if it won all its tenders, i.e. even if it priced its tenders at zero!

Invitation acceptance rate

Next we must consider the firm's invitation acceptance rate. This is the percentage of those projects on which it receives invitations for which it is actually willing to submit a tender other than a 'cover price' is the term for a tender priced in such a way that the firm making it knows that it cannot be succesful (be the lowest). That is, it is certain to 'cover' (exceed) the price of at least one other bidder. The firm is thus willing to produce should its tender be successful.

Why would a firm decline invitations to tender, or deliberately submit a cover price with no chance of winning? Leaving aside possible collusive market sharing behaviour as a motive for cover pricing, the reason would have to lie in the opportunity cost of scarce resources that, if committed to that project, would not be available for use on other projects, so that those projects could not then be tendered for or undertaken. At its minimum, this opportunity cost is the cost of preparing a tender. The firm's calculation may be that its probable profit from the project would be too low to justify incurring the cost of preparing the tender. Here probable profit is the best option available to the firm in terms of a set of trade-offs between each possible profit mark-up and the probability of winning the job with that mark-up. This is known as the 'expected value' of a bid.

Park (1966) and most subsequent authors take this probability rather than uncertainty approach to the problem. Thus, if the best available price option has, say, a 1 per cent chance of winning whilst making a 5 per cent profit margin, on a job with costs of £100,000, then the value of the right to bid is 1 per cent of £5,000, or £50. If it would cost more than this sum to prepare a tender, then bidding is best avoided. This is the best actual bid, however, in the sense that at any other price margin the profit multiplied by probability would have a value lower than £50.

Note that under many rules governing actual tendering, a tender bid, once made, cannot be withdrawn by the tenderer, without a penalty (loss of some form of deposit; and, certainly, loss of reputation). A firm can therefore risk losses if it tenders for jobs for which there is a high probability that it will not have the necessary available resources. If it does so, and wins the jobs, it is likely to face another kind of penalty. It risks penalties associated with late completion, since it takes time to acquire the needed resources and get them producing at peak efficiency.

Let us make the heroic assumption that there is a single known market price for construction output. Then, from the point of view of a contractor, the expected costs per unit of output could be lower on some projects than on others, although the value of each unit of output would be the same. A bigger gross operating profit, or contribution to overheads and profit, from another project might have to be foregone as the opportunity cost of taking a particular project. However, the firm may give up apparently lower unit cost projects because of intrinsic project characteristics. For example, a project with less full design may have greater scope for claims. Again from the contractor's point of view, some projects may gain from the effect of the passage of time. Later projects might be expected to be more profitable than today's projects on offer, because of expected changes in market price relative to the cost of production.

For realism, we must actually abandon the above assumption of pre-known market-clearing price. Each firm has to make its own guesses about present project price (price of lowest bid) and about future project prices. There is no futures market in construction reflecting the aggregate view of all participants about future prices, as there is for example in financial securities or agricultural commodities. Nor is re-trading or assignment of contracts allowed. Thus each contractor has to speculate as best it can.

The fact that project tenders are sequenced through time, and not simultaneous, is of considerable importance. If all projects that were going to start in year Y were put out for tenders on the first day of that year, contractors would face a quite different problem to the one they actually have to solve. If all projects were known at the beginning of the year, at the moment of pricing, contractors would know what aggregate and firm-specific demand actually would be for the period of production of those projects. They could be much more certain about actual opportunity cost. On the other hand, they would have only one chance to get their pricing right. There could be no iterative process of groping towards the most profitable attainable set of prices, using feedback information from the success or failure of previous tenders or continuously revised cost information from projects in hand. But mention of iteration and feedback is taking us too soon towards the *real* world. Let us return to this model world with its single market price for construction.

Tender success rate

The tender success rate is the percentage of projects tendered for, that a firm wins. The tender success rate is usually measured by value.

Assume a market clearing price exists, and this can be converted into a market price for each project. The tenderer's problem would then be to guess or estimate what this price could be, in terms of each project. Statistically if it were to exist, there would be one, and only one, tender equal to or lower than this price. A bid at this price would be the lowest bid. If there were likely to be two tender bids equal to or lower than this price, then it could not be called a market clearing price.

In a sense there would be a quantity of one project to be bought by the client, but offers of supply to do the project at that price would equal two. See Morishima (1984) on tender bidding as a reverse auction. We are left with a frustrated tenderer who was willing to sell the firm's services at that price, but was unsuccessful. Hence, the price was not an equilibrium market-clearing price, since by definition, at the equilibrium price desired purchases equal desired sales.

Each client may have an upper reserve price. If the lowest tender is above this, the client will then not award the project, but will postpone it or redesign it downwards in terms of quantity or quality. If there is no such reserve, the short run actual price elasticity of demand for construction projects will be zero. Aggregate short-run construction demand would be a vertical line, with the same quantity of work placed on the market regardless of actual tender prices. The market clearing price for a project obviously cannot be above the client's reserve price.

The problem is that each tenderer has to make a guess, and at least one of them is likely to err below the *true* price. What does this mean – after all if there is now a bid at that lower price, then surely that is now the market clearing price? Yes and no! The market clearing price has to be understood as the hypothetical lowest bid price that would obtain if all bidders were blessed with perfect knowledge in advance of actual costs and the market clearing rate of gross mark-up. In effect, all bidders would need a perfect knowledge of one another's intentions as well as of their own costs, including opportunity costs.

A better way of understanding things has been proposed by Fine (1975). Bidding is an inexact art. All bids are based *inter alia* upon estimates of prime cost. For any *n* bidders for the same project, these cost estimates will be spread over a range. If all bidders applied the same mark-up to estimated prime cost, each job would be won by the contractor who made the lowest estimate of costs. However, since each tenderer is honestly trying to produce as true an estimate as possible, it is statistically likely that the true cost lies somewhere towards the middle of the range of individual estimates. The lowest bid will have come from the firm that underestimated the true cost the most. This sets up a 'winner's curse'. The jobs a firm is most likely to win are those on which it has underestimated cost, and therefore over-estimated its out-turn profit. Thus *ex post* profit margins will be much lower on average than the *ex ante* margins firms think they have applied to their winning bids. The size of this effect increases with the average number of tenderers per project. Thus, if this effect is very strong, average *ex post* margins can even be negative. To protect themselves from this effect, therefore, firms add an error protection mark-up to their estimate, quite separate from and additional to their profit mark-up.

The more expert a firm's estimators are in the particular type of project being bid for, the lower the range of probable error, and the lower the error protection compensation mark-up that will be needed. Thus, firms with most experience of constructing the type of project in question will, *ceteris paribus*, be more likely to submit prices that are below the mean of bids received by the client, but less likely to submit the lowest bid, unless all bidders apply the policy of error-protection margins.

Since for each bid price there is a probabilistic range of possible outcomes, it follows that probability of success does not fall to zero if the firm raises its mark-up. There is always the chance that the bid will be the lowest. Success rate overall will fall if a firm raises its mark-ups on all bids. The question is, by how much will it fall? Suppose a firm receives invitations to bid for roughly £10 million of projects in a year, assuming the firm can put an approximate advance value on each project, so that it can be said to have £10m worth of invitations. Suppose in the past it has applied an average mark-up of 5%, and won on average 25% of what it tendered for. The result would be £2.5 million of jobs won, with a total profit of £125,000. Now, if it were to raise its mark-up from 5% to 10%, it would still increase its total profit so long as its success rate, and the value of orders won, fell by less than half, say, to £1.5m, or 15% of invitations.

Figure 10.1 is, therefore, the *meaningful demand diagram* for a contractor. The percentage mark-up is shown on the vertical axis, and success rate in value terms on the horizontal axis. The success rate in value terms is measured by the cost value of successful bids. R, T and V are combinations of percentage mark ups and likely tender success rates, taking into account the lumpy nature of construction. Given the

Figure 10.1 The gross profit-maximising strategy

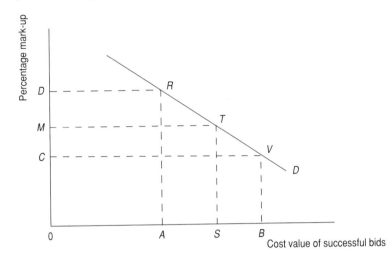

choice of R, T or V, the gross profit maximising strategy is found on the schedule linking trade-offs, at the point at which the area of the rectangle under the curve is a maximum, in this case at T.

It was assumed above that unit costs per project were unaffected by the number of projects performed. This simplified the analysis very considerably, but we may wish, for the sake of realism, to relax that assumption.

$$\text{Revenue} = (\text{cost of all projects successfully bid for}).\,(1 + M) \qquad (10.3)$$

where

M = the average mark-up.

Cost of projects successfully bid for is what was measured on the x-axis in Figure 10.1. Now, let us add a capacity-effect cost curve, whose height, measured as a percentage of normal cost, rises in some known way as output increases beyond a least-cost or economists' capacity level. Thus

$$C = c \,.\, (S - S_e) \qquad (10.4)$$

where

C = total cost of working above capacity
c = the slope of the capacity effect schedule
S = the actual level of output

and

S_e = optimum level of output in terms of unit cost

Adjusted for working beyond most-efficient capacity, therefore, *ex post* profit margins fall in proportion as the capacity-effect rises.

We now have in place the fundamentals of a model of how individual bidders determine their price for a single project. The next step is to bring together the *n* bids thus generated, to determine the value of the lowest, and thus the contract price.

Price determination for a single contracting project

Now, in the UK as in most countries, construction contracts are awarded by a reverse or seller's auction process, in which each bidder is allowed one bid and the project is awarded to the lowest bidder. Moreover, this is a single, sealed-bid form of English, not Dutch, auction. In an English auction there are many bids. In a Dutch auction the transaction is awarded to the first bidder to accept the auctioneer's proposed price, so that only one bid is ever uttered. If Dutch auction rules were followed in construction bidding, a client would begin by announcing the low price they proposed to pay a contractor. If just one supplier was happy to accept this price, a deal would be done. If no-one bid, the client would offer a revised, higher price; and so on until there was a bidder. There is indeed anecdotal reportage that Dutch auctions are used by some main contractors in appointing subcontractors. Unfortunately, we possess no systematic study of actual subcontract procurement practices – a remarkable absence, considering the multiplicity of studies of main contract procurement practices!

In some countries, clients take expert advice from the local equivalent of quantity surveyors as to what offer price is likely to attract one willing bidder, and announce this as the project price in advance of inviting bids. If the price announced is too high, so that there is more than one willing bidder, the auction proceeds by inviting bidders to offer a percentage deduction below the announced price. The contract is then awarded to the bidder offering the biggest percentage deduction. This method is common in southern Europe, where it seems to have been developed to assist tendering by technically inexpert contractors in need of help in identifying the likely construction cost of a project.

However, aside from this local peculiarity, no *price* for a construction project can be said to exist before firms decide indirectly what that price is going to be, by directly determining their own bid price. How can such bidders be said to be *price-takers*?

Moreover, each heterogeneous construction project has its own specific price; and, unless we have schedule-of-rates contracts, it is not an easy matter to convert contract values into prices for homogeneous 'units of output', as in a demand-and-supply diagram. Prices of elements or items in Bills of Quantities will not do for this purpose, because they are not market prices, and do not constitute subtractable offers to perform an item of work in isolation from the whole package. Actual construction bid prices are arrived at by each bidder estimating costs (perhaps only direct, variable costs) and then adding a mark-up to that estimate, to cover fixed, overhead costs plus profit.

It is certainly true that cost estimating is prone to error. Given the same documents to cost, different estimators will come up with different figures, even if asked to estimate direct cost only. These divergencies can be described as errors if we introduce the further assumption that, in the out-turn all bidding firms would actually have experienced the same level of cost. Thus a *true* cost for the project is said to exist, and each firm's estimate is an error-prone attempt to calculate this true cost, in conditions of incomplete information and bounded rationality.

But it is by no means clear that such divergencies derived from error are the main reason for differences between bids for the same project. One possibility is that different firms face different real costs of constructing the project, either because they use different technologies or differ in the efficiency with which they use a technology. A second, perhaps more important, possibility is that different firms add different mark-ups. A market clearing price from the contractor's point of view is a contract-winning price, which is only marginally below the next lowest bid, and yet which gives the winning contractor a profit margin consistent with market equilibrium.

We begin by assuming firms have some, though perhaps imperfect, advance knowledge of the identity of rival bidders for each project, and of those rivals' recent tender record. Now, assume that each firm first calculates its *minimum acceptable price* (MAP) for a project. The firm's MAP would be the lowest price at which it would be very surprised to make a significant out-turn loss, and at which opportunity costs of winning a project are covered. It then determines by how much its actual bid price should exceed this minimum. In recessionary conditions, the opportunity cost of winning the project may be negative, because there is no alternative mutually exclusive use for internal resources, and there would follow a loss of positive cash flow if the project were lost. In boom conditions opportunity cost will be substantially positive. If the firm is falling short of its target turnover, its bid price may therefore be below its estimate of an avoidable project cost recovery price. On the other hand, in strong demand conditions, opportunity costs may very substantially raise its minimum acceptable price.

If probability comes into the picture at all, it is via estimation of the subjective probabilities of winning the project associated with each level of excess over the MAP, based on knowledge about rival bidders and the number of bidders. We treat the determination of the firm's bid price as analogous to the bilateral monopoly problem first analysed by Edgeworth, who showed that the solution was indeterminate, depending finally upon bargaining skills and subjective assessment of uncertainty, within determined upper and lower limits – the seller's MinAP and the buyer's MaxAP. The fact that there are rival bidders is treated as one of the areas of uncertainty facing the seller's 'negotiator'.

However, the model must now be modified to take account of the fact that bidders may expect the out-turn price to exceed the tender price, and that this will be taken into account in calculating the MAP.

Tender and out-turn prices and economic theory

Our present concern is with contractors, for whom it might well be thought that the gap between expected and actual price is a non-issue, since by definition for each project a contractor agrees a selling price with a client before they start production. However, things are not so simple, mainly because out-turn price received by a contractor can differ widely from tender price. Final prices may be greater than tender prices because of claims for variations or delays largely to do with incomplete pre-tender design, opportunistic behaviour by contractors and clients, and information-impactedness.

Prices are at the centre of economic theory because economists assume that, in one way or another (and they differ on which way it might be) prices are the mechanism of resource allocation. Prices are at the centre of the attention of buyers and sellers, customers and firms, on the other hand, because they are important influences on value-for-money, revenue and profits.

These are different reasons for being concerned with price, and in fact they refer to different concepts of price. For buyers it is *ex post*, out-turn prices that affect value-for-money. For sellers, it is both kinds of price that matter, because *ex ante* prices affect their volume of contracts won or sales, whereas *ex post* prices affect their profit per unit of sales. For economists, it is *ex ante*, tender prices that matter in the short run, because they affect immediate resource allocation – for example, the allocation of demand and output and resources between rival producers, perhaps with different levels of efficiency – though if in the long run retained profits of a firm affect its investment, then *ex post* prices too are important for long run resource allocation. One of the things that makes construction prices a fascinating, though frustratingly complex, subject of study is the 'stylised fact' that in construction these two concepts of price rarely coincide.

What can economics say about the divergence? In other words, can 'claims and variation orders' be made into a variable in an economic model? The branch of economic theory making most progress in this direction is surely transaction cost theory (see Chapter 5).

We would also argue the necessity of basing economic models in real, historical time, to grasp the fact that price normally responds to demand in two staged processes. First, a change in demand affects tender prices. Then, later, it has further effects on out-turn prices. It would be illuminating to be able to study the movement in the ratio of out-turn to tender prices (for 'fixed price' contracts) over the successive stages of a demand cycle. Unfortunately, this has not yet been attempted.

Contractors' pricing and target output levels

All the above arguments together offer an account of how contract construction price levels move through time in response to changes in demand conditions. Where there is a market comprising clients on the one side and main contractors on the other, prices adjust to reflect imbalance between construction firms' target levels of turnover and actual levels being achieved at the former price level.

In recessions firms may fall short of their target turnover, and react by cutting their tender prices. If there are a sufficient number of such firms, this fall in prices spreads to rivals and becomes the new general price level. This, however, does not constitute an equilibrium, because the price elasticity of aggregate construction demand is low. Output volumes increase a little, but the effect of this on profits and turnover will be more than offset by the fall in price per unit. Therefore, once the lower prices become general, all firms find themselves falling even further short of their target turnover.

Moreover, lower main contractor prices may result in a downward movement in their direct costs, as measured by subcontractors' prices. This encourages some main contractors to attempt a further round of price cutting, and so the process continues. The magnitude of the effect depends on the size of the target turnover shortfall. This shortfall depends on the overall size of the drop in demand, on the rate of change in the number of firms in the market, and on the firms' non-price competitive position. Lower main contractor prices also depend on clients' and main contractors' procurement practices. In particular, the greater the length of tender lists, the greater the size and duration of the downward dynamic effect on prices. The fall only ends when firms reduce their target turnovers sufficiently, or the number of firms falls sufficiently, to eliminate the deficit for each firm between actual and target turnover.

In periods of economic growth, firms may exceed their target turnover, and react by raising their margins and tender prices. Once sufficient firms do this, the rise in prices spreads and becomes general. In one version or account, this process begins with subcontractors. Main contractors then pass on these higher subcontract prices in their own tender prices, and add a wider margin of their own. If labour costs and materials prices are also rising, this strengthens the upward movement of prices. Thus, in periods of expansion though not in recessions, according to this account, main contractors tend to *take* subcontractors' prices as setting a given level of cost.

Alternatively, the process may begin with main contractors raising their prices. This then has the effect of raising the maximum prices they are prepared to accept from subcontractors. This, together with subcontractors raising the minimum prices they are prepared to accept, raises upward the whole Edgeworth-range of indeterminacy of negotiated prices. Subcontract margins and prices then follow main contract margins and prices upward.

Again, since aggregate price elasticity of demand for construction is generally quite low, firms find that they are still exceeding their original target turnover. Now, if many construction firms respond to the boom by investing in ways to increase their capacity, and hence raise their turnover targets, the price rise may now subside. However, if firms do not raise their capacity, they cannot raise their target turnover, unless they raise their prices. Another round of price increases is therefore likely. Again, client and main contractor procurement practices and the number of existing firms and new firms entering the industry affect the process, via their effect on the suppliers' opportunity costs and hence on suppliers' minimum prices.

Concluding remarks

The key features of the model of contractors' pricing strategies given in this chapter are:

- the use of the concept of target turnover, for both current and future production periods.
- the treatment of construction pricing for each project as sequential, with consequent scope both for 'regret' and for adjustment.
- the absence of equilibrium, or of a role for the concept of marginal cost.
- the treatment of competitive tendering project price outcomes as indeterminate, within a bounded range.
- the assumption of highly variable *ex post* margins.

This account therefore differs from conventional ones, even of the price-setting or cost-plus family.

It is the beginnings of a theory which we hope, when developed, will be able to explain and understand rather than to make predictions. We believe that it has the potential to explain, at least as well as any other theoretical approach in the field of construction economics, the observed behaviour of individual firms and the observed movement of the general level of construction prices, relative to both demand and cost, during construction demand cycles in the UK.

Part V
Capital Circuits

11
The Industrial Capital Circuit and Cash Flows in Construction Firms

Introduction

This chapter begins with a discussion about the economic implications of management in a firm. We look at the organisational and accounting concepts used to define and measure a firm. We then go on to discuss the industrial capital circuit and the social structures of accumulation, which form the preconditions necessary for capital accumulation by firms to take place. This approach using capital circuits enables both organisational and accounting approaches to be used in the same model. We then compare and contrast profits and cash flows.

Organisational and accounting concepts of the firm

Penrose (1980) defines a firm essentially as an organisational entity, thus:

> It is the 'area of co-ordination' – the area of 'authoritative communication' – which must define the boundaries of the firm for our purposes ... The concept of the firm developed above does not depend on ... the mere existence of the power to control ... If a corporation is controlled by, or because of stock ownership is classed as a subsidiary of, a larger corporation, it is part of the larger firm only if there is evidence of an administrative co-ordination of the activities of the two corporations ... It should not be classed as part of the larger firm if it appears to operate independently of the managerial plans and administrative arrangements of the larger firm, for in this case any influence the larger firm exerts should be viewed as an

extension of economic power and not as an extension of the co-ordinated planning of productive activity. (Penrose, 1980, pp. 20–1)

She also offers clear guidance on the problem of defining the continuity of existence of a firm:

> In practice the name of a firm may change, its managing personnel and its owners may change, its geographical location may change, its legal form may change, and still in the ordinary course of events we would consider it to be the same firm ... Whether the continuity was maintained by bankers in times of crisis or by the ingenuity of a clever (company) promoter is irrelevant, provided that the firm neither suffered such complete disruption that it lost the 'hard core' of its operating personnel, nor lost its identity in that of another firm. (Ibid., 1980, pp. 22–3)

In contrast to the *organisational* approach adopted by Penrose is the definition of the firm in terms of what it *owns* and of *who owns it*. This ownership approach is used in accountancy and by the stock market. The firm owns various kinds of assets. When measured net of liabilities, these assets constitute 'what the firm is', and this can be measured as a number, a value of net assets. This in turn is what the shareholders own. The firm as defined in purely legal terms, exists only so long as it continues as a legal entity, owning positive net assets, and, in ownership terms, as long as it is not the property (subsidiary) of another firm.

Now, in Penrose's sense what is striking about the larger construction contracting firms has been their relative longevity. Certainly this has been so when we compare them with other organisations of the modern construction industry such as professional service firms or speculative builders. Financial failure and consequent changes in the owners of large contractors is relatively common, but it is less common for financial problems to be allowed to end the existence of such firms. The chief intangible asset of any contractor is not any product brand, but the name and supposed longevity and consequent soundness of the firm itself. The construction world is in some ways very conservative, and one manifestation of this has been the reluctance of new corporate owners of construction firms to merge their organisational identity with that of others or to abandon use of the pre-corporate name.

Thus, subsidiaries of the former Trafalgar House such as Cementation or Trollope & Colls remained organisationally and nominally distinct entities even decades after their take-over. On the other hand, Kvaerner has now converted Trafalgar House Construction to its own corporate name. Old-established civil engineering contractors such as Kier or Nuttall have endured through many changes of ownership.

In the terms of the construction industry, Tarmac Construction Ltd (now Carillion) still counts as a relative newcomer and young organisation, compared with Taylor Woodrow, Laing, Mowlem, Costain or McAlpine. All this longevity, moreover, has been achieved in a business rightly characterised as one with low barriers to entry.

Although they may have survived as organisations, ownership of the larger construction companies has in recent decades begun to be much less stable. Even twenty-five years ago, the *family firm* was the dominant ownership-type in the British construction industry. Such family firms bore the name of their founders, and key boardroom positions and controlling blocks of shares were still held by descendants or the representatives of family trusts. Often the venerable founder, or their second-generation successor, remained as company president. For the most part, the age of opportunity to establish such business dynasties lay in the past, but still in living memory, namely the period of the speculative housing boom of the 1930s, the Second World War and its post-war aftermath (Smyth, 1985).

Firms that had achieved a leading position by 1950 had need for little in the way of new-blood innovative entrepreneurship in order to retain their position in the market through the following twenty-year *golden age* for UK construction. The leaders and business practices that had made them strong in the first place remained sufficient to keep them so.

Today, a remaining successful family firm, such as John Laing plc, is much more unusual. The majority of shares in most cases have passed to the ownership of financial institutions. Company boards live in some fear of the verdict that fund-managers and city analysts will pass on the quality of their leadership, strategic vision and financial performance. The boards of large construction firms are increasingly composed of company people who over a working lifetime have risen from technical or professional positions as employees in the company. The boards also have increasing numbers of 'financial engineers', who are specialists in the raising of corporate finance, the arrangement of financial deals, and the management of relationships with the City. Usually these board members with a financial expertise are persons without a lifetime background in the construction industry, still less the firm itself.

No construction firm is now too large to be swallowed by an acquirer. A low share price can make virtually any firm vulnerable to take-over – if, that is, anyone wishes to buy it! Low share prices may result from unsuccessful property or housing development activity, or failed attempts to diversify by acquisitions outside construction, or perhaps, simply be the consequence of a long period of unprofitable trading in a firm's core contract construction markets.

To give just one example, during the early 1990s Davy, one of the leaders in certain branches of the world market for construction engineering, was acquired in 1991 by Trafalgar House, at that time under the control of a *financial engineer* and entrepreneur, Sir Nigel Broakes, who had founded Trafalgar House, initially as a property company, later diversifying it into construction and shipping amongst other industries. It became a leading UK example of the conglomerate type of corporation. Next, a controlling stake in Trafalgar House was acquired by the Jardine Matheson group, replacing Broakes and his former chief directors with representatives of the new owners. Then, in 1996, the whole ownership of Trafalgar House passed, through takeover, to a Norwegian engineering and shipbuilding group, Kvaerner.

Thus the operating companies formerly in the Davy group, such as Davy McKee (London) Ltd had undergone three changes in the identity of their owners in less than five years. However, whereas the acquisition of Trafalgar House by Jardines was only a change at the financial and corporate-strategic level, acquisition of Davy by Trafalgar House saw certain Davy-owned businesses, such as Davy Offshore, lose all or much of their organisational separate identity, being fused into existing Trafalgar House-owned businesses operating in the same fields, or rationalised out of existence, as redundant duplication.

Measuring the size of a firm

Following Penrose in defining the firm as an organisation controlling, by administrative decision, the use of a collection of productive resources (Penrose, 1980, p. 24), it follows that, ideally, we would agree that 'the size of the firm is best gauged by some measure of the productive resources it employs' (ibid., p. 24).

Note though that the firm does not own many of the key resources, the co-ordination and control of which define it as an organisation. This is, of course, particularly true of its staff. On the other hand, it may well own various assets which are not key resources, in Penrose's sense. It may, for example, hold shares in other firms.

There are, therefore, serious difficulties involved in measuring resources. In particular, it may even be inappropriate to use the balance sheet valuation of assets as a measure of a firm's non-human physical resources. There is an analogy with the distinction in macro-economic national accounting between Gross Capital Stock and Net Capital Stock. Gross Capital Stock measures the productive resource constituted by a stock of durable goods, such as machines, vehicles, and buildings, whereas Net Capital Stock measures their market value, and therefore the wealth of their owners. If company accounting attempts to measure productive resources, balance sheet valuations may be a poor guide to their market value.

As economists, we are concerned with the firm in two respects. First, the firm may be seen as an institution through which productive resources are directed, allocated and co-ordinated. In this respect, we are concerned with efficiency in the use of resources. Second, the firm may be seen as a legal ownership entity. This aspect is of interest to economists, because of our concern with the distribution of wealth (and economic power), and because, after all, it is capitalism that we are trying to understand. In what follows, therefore, we shall have no choice but to move to and fro between these two ways of looking at the firm, corresponding as they do to the perspectives of managers and owners.

Finally, for some purposes, either turnover (sales; gross output) or value added are preferable to capital assets as measures of a firm's size. Turnover may be of most interest if we wish to place a business in the context of the overall market for its product, and ask, what share of that market does the business command? Value added may be of most interest if we wish to measure the contribution of a firm's activities to the national product and national income, or to see whether a firm's output is growing in real terms, or to compare output with resource input, and hence to measure productivity, or its change. Unfortunately, individual UK firms do not publish data in value added terms as part of their company accounts, though it may be possible to arrive at an approximation by adding back the UK wage and salary bill, often reported, to the figure for gross profits.

Thus, company size may theoretically be measured in terms of physical plus human resources, the capital or *net assets* belonging to the firm, the firm's *turnover*, or its value added. Of these only its net assets and turnover are readily measurable from the data as supplied in company accounts.

Capital circuits: a possible synthesis of the ownership-based and resource-based approaches

In principle, speculative developers receive a lump sum on completion and sale of a building. In practice they may arrange for finance or deposits to ease their cash flows, but these are all liabilities and returnable until completion. Contractors and subcontractors, and professional practices, mostly rely upon receiving the full payment for the value of their output either in monthly or other timed instalments, as production proceeds. From a production perspective, contractors can be regarded as receiving credit from their clients, in that they receive part payment before production of the project is completed. Each interim payment for part of the work on a building or civil works is non-returnable although the project is incomplete. However, from a sales perspective, the clients can be seen as receiving trade credit from the contractor, since all the works have been sold the moment the contract is signed, though they are only paid for in lagged instalments later.

Construction firms also rely upon receiving credit, called *trade credit*, from their suppliers, who may be manufacturers, builders' merchants or subcontractors. They obtain trade credit because they pay for goods and services from these suppliers some time after they are received. Together, pre-payment by customers and trade credit by suppliers greatly reduce the amount of working capital of its own that a firm needs in order to finance its operations. Without pre-payment and trade credit each construction firm would only be able to finance the production of a much smaller output or turnover than it actually achieves. Otherwise the firm would have to increase either its share capital or its bank debt or both.

Capital circuits and company accounts

A company's balance sheet is drawn up at one moment of time. Balance sheets therefore take a still photograph of what is a continuous process of buying, using, paying, making, selling and getting paid, recording the condition of all inputs and outputs as at that moment. This continuous process can be analysed using an alternative theoretical approach, called capital circuits, and an alternative method of measurements called cash flow statements. Assume for the moment that inputs are paid for when or before they are received, and that receipts of money from the sale of outputs are obtained after the output has been produced and sold. The circuit of circulating capital for a producer, ignoring fixed capital for the present, is then as shown in Figure 11.1.

In Figure 11.1, M is money, C represents any commodity and P is the production process. Time is shown by movement from left to right through the circuit. The firm begins with a sum of money or bank credit. This money is used to purchase inputs in the form of commodities (materials and components). These inputs, together with labour, are then used in a process of production in which they disappear, but out of which emerges a new product, C'. This is then sold, and payment of money received. With this money, the circuit is then restarted.

At any snapshot moment of time, then, part of the firm's circulating or working capital will take the form of M, money received from previous sales and not yet spent on purchasing inputs; part will take form C, commodities bought and taken possession of but not yet used; part C', outputs produced but not yet sold or paid for; and part will take form P, work in process of production. Some circulating capital will take the form of debts or credits, to counter-balance inputs received and valued as part of C but not yet paid for, or outputs sold but for which payment has not yet been received. This is the moment-in-time consequence of the cash flow cycle through which capital revolves.

Any construction firm obtaining contracts to supply its goods or services to a buyer can be said to have sold its output at the moment the contract is signed, even before production has yet begun. However, payment will only be received later, usually in instalments as parts or periods of the production are completed. Thus a firm would be entitled to report the total contract value of projects won in a year as the value of its sales in that year, although much of the production of those projects would not occur until after that year was over.

In order to match production with sales, it is normal accounting practice in construction businesses not to measure turnover or sales at the time of signing a contract. Instead, output is counted as *turnover*

Figure 11.1   The simple industrial capital circuit

Time

only when it has been produced and, usually, valued by the customer's representative, probably according to unit prices for elements of work in a Bill of Quantities. From the point when an interim valuation and certificate is signed, the output is counted as sold, but, until the customer has actually paid for it, it appears as *trade debt* in the firm's accounts. There is also a period between work being done and valued. Output in this condition is called *work in progress* in a firm's accounts. Finally, some inputs of materials and components may be purchased and delivered into the possession of a firm some time before they are to be incorporated into the project in hand. The value of these items is counted as *stocks* in the firm's accounts.

The main feature of a capital circuit is that capital or money is used by a firm or business to create increased amounts of capital. In other words, capital is used to accumulate capital. Indeed the process of capital accumulation becomes an end in itself and gives the economic system its character. Kotz *et al.* (1994) propose a useful grouping of conditions (external to the firm itself) necessary for capital accumulation to take place. These conditions form a social structure of accumulation (see Chapter 1). The industrial capital circuit depends on:

- conditions of input availability, assuring the $M \rightarrow C$ part of the circuit. The necessary inputs into the production process must be available on the market to be purchased by capitalists with the necessary money;
- conditions of production, assuring the $C \rightarrow P \rightarrow C'$ part of the circuit. Input-output coefficients must be reasonably endogenously stable, and production must not be disrupted by exogenous factors. Above all, this means that the labour process is under some degree of stable capitalist control;
- conditions of realisation, assuring the $C' \rightarrow M'$ part of the circuit. Output produced must find buyers, who must possess means of payment, and there must be no major collapse of effective demand;
- conditions of investment financing, assuring the reflux of M' from the end of one circuit back to finance the initiation of another circuit.

The industrial capital circuit is not the only way capital can be used to accumulate more capital. Robinson and Eatwell (1973) discuss three distinct ways of using money to make money. They are the productive circuit of industrialists, the commercial circuit of merchants and the financial circuit of bankers. To these three capital circuits can be added a fourth, which is of relevance to the construction sector, namely, the property circuit of the owner of real estate. We shall only discuss the industrial circuit in this chapter. The other three circuits will be discussed in Chapter 12.

The industrial (productive) capital circuit

Production is carried out by industrial capital. Let us develop the industrial capital circuit shown in Figure 11.1. Figure 11.2 is also a schematic representation of the industrial or productive capital circuit. As before, an industrial producer uses an amount of money capital to purchase commodities, now of three separate kinds, namely raw materials or manufactured inputs, labour power, and durable means of production (plant and machinery; buildings and works). These are combined in the production process to create finished products or services. A production process adds value to the commodity inputs as they become finished products which can be sold for an amount of money, M', which is greater than the total expenditures on production, $C + LP + MP = M$. In other words M' is greater than M and the difference $M' - M$ is the profit of the firm. The circuit is completed when M is returned by the firm to purchase a further quantity of materials and re-hire labour to start a new round in the cycle of production.

Figure 11.2   The industrial capital circuit, showing labour, means of production and net operating cash flow

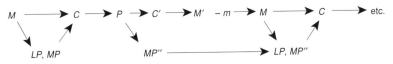

where

$M =$ an amount of money capital

$C =$ commodity inputs, materials (resource), or value of stocks (accounting)

$LP =$ labour power, the productive power of labour (resource), or the part of added value paid as wages (accounting)

$MP =$ the productive power of means of production (resource), or the value of investment in fixed capital (accounting)

$P =$ the productive services of resources in the production process (resource), or the value of work in progress (accounting)

$C' =$ output from the production process (resource), or value of finished products not yet sold plus trade debtors (accounting)

$M' =$ an amount of capital greater than M

$m =$ $M' - M =$ net operating cash flow

$MP'=$ the remaining productive power of the non-consumed part of the means of production (resource), or depreciated value of fixed capital assets (accounting).

For example (and simplifying so that overhead costs consist only of the costs of fixed capital, MP):

Figure 11.3   Numerical example of the industrial (productive) capital circuit

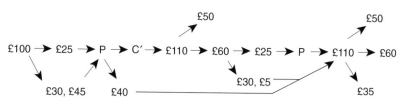

In the industrial circuit shown in Figure 11.3, for every £100 of finance spent at the setting-up of production, £25 is used to purchase raw materials, £30 to pay wages and £45 to purchase new machines. Combining these resources constitutes the production process (P) out of which emerge the finished goods or output (C'). The output is then sold for £110, realising £10 more than the original money capital used.

The net cash flow in the first period is the difference between the gross revenue, £110, and the total production cost, including investment expenditures, £100. The net cash flow therefore measures the addition to the stock of money obtained in the first period by the industrial capitalist. In the example in Figure 11.3, the economic surplus created by the production process might be thought to be £5 less than the net cash flow, because £5 is the amount that must be set aside to replace the part of the stock of fixed capital used up in the production process. On the other hand in the next period, an expenditure of only £60 (£25 on materials, £30 on wages and £5 on fixed capital investment, to restore the fixed capital stock) generates a gross revenue of £110, and therefore a net cash flow in the second circuit of £50. The annual sustainable economic surplus resulting from this production process is in fact £50, which is the excess of the value of output over the value of inputs used up in order to produce it.

We have defined money capital accumulation (the increase in the stock of money capital belonging to the capitalist) so that it has the character of a cash flow concept – it relates to expenditures made and revenues received during the period. Whereas, we have defined economic surplus in 'resource' terms. It is the money-equivalent of the extra economic resource potentially made available, either for an increase in consumption or an increase in the resource-stock, as a result of production – the excess of outputs produced over inputs used up.

We now need to compare with two concepts introduced in earlier chapters, those of gross operating surplus and net operating surplus. Gross operating surplus is a cash flow concept. It measures gross revenue minus actual expenditures on production, but excludes investment expenditures. Net operating surplus is *not* a cash flow concept, and is measured by deducting depreciation from gross operating surplus. This depreciation does not measure an actual expenditure, or cash flow. Its deduction to arrive at net operating surplus is thought appropriate because net operating surplus is a measure of the amount available to be appropriated by bondholders, banks, government and shareholders. If no allowance for depreciation were made, the surplus actually sustainably available for appropriation would be overstated. Neither of these two concepts deducts actual investment expenditures, however, from gross revenues.

We need these concepts because, in fact, investment costs and production costs are decided differently, and are treated differently in firms' accounts. Investment expenditure is, in the short term, discretionary or avoidable, because whatever the level of investment spending, production can go on. Production costs, on the other hand, are sometimes known as *costs of sales*, because any given level of sales revenue requires a certain level of production expenditure. Gross operating surplus is therefore the maximum amount of cash that can be taken out of a business whilst it continues to operate in the short run at its original level.

Net operating surplus is sometimes called net profit or 'profit before interest and tax'. However, we will reserve the term 'net profit' for 'profit after interest and tax', a concept also known as 'profit available to shareholders'. Actual investment expenditures are *regarded* by firms and others as financial from their retained cash flow, which is retained net profits plus depreciation provisions.

Hence:

Net operating surplus = Gross revenue – Costs of production and administration – Depreciation.

Net profit = Net operating Surplus – (Interest + Rent) – Tax.

Retained profit = Net profit – Dividends.

While:

Economic surplus = Gross value of output – Value of inputs used up in its production, including consumption of fixed capital.

Whereas net operating surplus is struck after deducting total production and administration costs, the economic surplus is theoretically calculated after deducting only the amount *necessary* to replace inputs used up in production. Thus, if inputs are purchased at 'above their value', or if unnecessary inputs are purchased, net operating surplus will be below the size of the economic surplus. Both these things may occur. Employees may obtain their own share in the economic surplus generated by a firm, by obtaining wages above the market price of their labour power, charging an economic rent on their labour. Indeed, in successful firms generating significant economic surpluses, the conflict between the owners of the firm and its workers over their respective claims in and shares of the economic surplus generated by the firm's production may be a major and recurring phenomenon. At the same time, the managerial staff apparatus and their expenditures may be deliberately or otherwise inflated beyond its *necessary* size, thus also claiming a share of the firm's economic surplus. These expenditures may be purely wasteful, such as the perquisites of a controlling though non-owning stratum of managers, or they may, like investment expenditure, be future-orientated. Thus expenditures incurred in building up a company's image, name, market-presence, and brands are not necessary costs of the present circuit of production. However, they may operate analogously to investment expenditure in increasing the firm's economic surplus and, in particular, its gross operating surplus, in the future.

Perhaps because major marketing expenditures of this kind are a relatively recent phenomenon of the twentieth century, they are not treated in firms' accounts on a par with investment in fixed capital or circulating capital stocks. Instead they tend to be mixed in with production costs under a generic term such as operating costs. Intangible assets do often appear in balance sheets, but expenditure on building up the value of these intangibles is more rarely measured as investment expenditure, especially in construction related businesses.

The business and financing cash flows of an industrial firm

In the simple example of Figure 11.3 there is neither expansion of output nor borrowing by the firm. Thus there are no interest payments.

Net profit is then equal to net operating surplus, less tax. This meant that the whole after-tax net operating surplus was treated as available either to be distributed as dividends to its owners or retained to finance investment, and as belonging to those owners and adding to their wealth.

But, suppose that the firm obtained its original £100 of money capital partly by issuing shares to the value of £50, and partly by borrowing £50 for 5 years on a long-term loan. Suppose further that the interest rate, i, is 5 per cent. We would now have to deduct interest payable on that loan from the net operating surplus to arrive at the net profit attributable to shareholders. Interest payable by the firm would be £2.50, whilst repayment of the principal would fall only at the end of the loan.

Net cash flow was defined above as gross revenue less gross expenditures *including* investment expenditures, assuming no borrowing. Now we shall have to modify and clarify this. Part of investment is financed out of retained profits. The whole of investment expenditures financed this way constitutes a net cash outflow at the point when the investment is made. But part is financed by borrowing. In this case, first, there is a cash inflow, which is the sum borrowed, but this is equal to the outflow of that part of the investment expenditure paid for by borrowing. Thus borrowing to invest has no immediate impact on the net cash flow. However, subsequently there will be a stream of interest payments, which are cash outflows; and finally, the principal of the sum borrowed will have to be repaid – another outflow.

The net cash flow profile would then be as illustrated in Table 11.1, assuming the firm commences business with £50 of its own capital, which is its shareholders' funds. Net cash flows then measure the amount by which this money capital increases:

Table 11.1   Net cash flow table

Year	1 £	2 £	3 £	4 £	5 £	6 £	7 £	8 £	9 £	10 £
Revenue	110	110	110	110	110	110	110	110	110	110
Loan	50									
Total in	160	110	110	110	110	110	110	110	110	110
Less										
Production cost	55	55	55	55	55	55	55	55	55	55
Investment cost	45								45	
Interest	2.50	2.50	2.50	2.50	2.50					
Loan repayment					50					
Total out	102.50	57.50	57.50	57.50	107.50	55	55	55	100	55
Net cash flow	57.50	52.50	52.50	52.50	2.50	55	55	55	10	55

Net cash flows are the composite result of what we may call *business* cash flows and *financing* cash flows. In Table 11.1, year 1, for example, the business cash flow consists of £110 inflow of operating revenue, and a cash outflow of £100 of production costs plus investment costs, a net business cash flow of £10. In the same year the *financing* cash flow consists of an inflow from a loan of £50 and an outflow of £2.50 in interest payments, a net financing flow of £47.50. Whereas the firm began the year with cash of £50 (its share capital), it ended it with cash of £107.50, consisting of its original £50 plus the £10 of net business cash flow and the £47.50 of net financing cash flow.

Business cash flows are determined by the nature of the business being conducted by a firm, which determines its revenue from sales, its production costs and its need for investment in order to undertake that production. They are independent of how the firm raises its finance in order to undertake this business. The nature of the business determines the total amount of finance required to carry out a certain volume of production.

In our example, we made two assumptions concerning cash flows. Firstly, it was assumed that all purchases of inputs had to be made for cash in advance, and had to be made at the start of the year. This is why the firm needed an initial money capital equal to its total annual expenditure. Secondly, the time lag from purchase of inputs to sale of output was assumed to be exactly 12 months, so that revenue flowed in only at the end of the year. If some revenue had come in earlier, then it could have been deposited in a bank to obtain interest, but our simplified accounts show no such thing. But if revenue from production and sales in the first year had come in after the first year, then it would not have been included in that year's cash flow, and would not have been available to finance the following year's production costs.

Financing cash flows, on the other hand, are determined by the firm's choice of how to raise finance, as between debt and equity. Use of debt implies interest and repayment outflows, set against the initial inflow of the sum borrowed. Issuing new share capital of the same nominal value brings an equivalent initial inflow, and implies dividend outflow. In our example, the firm chose not to expand, and thus had no need for retained profits. In years 2 to 8 it needed only £55 of operating capital, sufficient to pay production costs. The entire accumulated cash stock over and above this could have been distributed as dividends. In our highly unlikely example, the firm is so profitable that it need not even set aside part of annual net cash flows into a *reserve* to be used to meet future liabilities when they arise, such as the repayment of the loan or

Table 11.2   Net profit table

Year	1	2 £	3 £	4 £	5 £	6 £	7 £	8 £	9 £	10 £
Revenue	110	110	110	110	110	110	110	110	110	110
Less										
Production cost	55	55	55	55	55	55	55	55	55	55
Depreciation	5	5	5	5	5	5	5	5	5	5
Interest	2.50	2.50	2.50	2.50	2.50					
Net profit	47.50	47.50	47.50	47.50	47.50	50	50	50	50	50

the replacement of fixed capital equipment. Instead, these liabilities can be met simply by reducing that year's dividend, and diverting the net operating cash flow temporarily to that purpose.

Meanwhile the net profit profile would be as shown in Table 11.2. Note that, at the end of year 5, the £50 loan principal would have been repaid. However, loan repayments are not normally lumped with interest, and deducted from net operating surplus before measuring net profit in the profit-and-loss account. Instead, the loan repayment appears as a liability on the balance sheet. When the loan is about to become due for repayment, its amount is deducted from *long term liabilities* and added to a balance sheet item called current liabilities. Let us assume that, at the end of each of years 1 to 5, not all of the net profits were distributed as dividend, but that instead £10 was added to the balance sheet as profits transferred to reserves. This annual transfer to reserves would therefore accrue to £50 by end of year 5, from which the loan could be repaid. This then simultaneously removes £50 both from the asset and the liability sides of the balance sheet.

Hence, ignoring for the moment the existence of company taxation:

Net profit = Net operating surplus – (Interest on loans + Rent)

and

Retained profits = Net profit – Dividends

and

Net worth = Initial share capital + (Σ Retained profits) – Long-term loan repayment liabilities

From these expressions it can be seen that net profit is the residual left after interest and rent have been deducted from net operating surplus.

Retained profit is the residual left after dividends have been deducted from net profit. Similarly net worth is the residual left after long term loans have been deducted from the share capital and accumulated retained profit.

Measures of the economic performance of an industrial firm: economic surplus, profit and net cash flow

- Economic surplus is defined by deducting from the value of output the amount needed to maintain a constant productive capacity and output in the next period.
- The net operating surplus is defined by deducting from revenue the costs of production and administration and an allowance for depreciation.
- The business net cash flow is defined by deducting from revenue the costs of production and administration as well as investment expenditure.
- The financing net cash flow is defined by deducting from new cash sums raised as equity or debt, if any, the amounts of interest payments and dividends and repayment of loans.
- The retained profit is defined by deducting from net operating surplus all the amounts actually disbursed as interest, tax and dividends. However, it is not a measure of cash flow, because net operating surplus is calculated after depreciation provision.

Some of the net operating surplus is used to pay interest to the owners of part of the capital, M, used at the outset of the process. Either capital is borrowed or the owners of the firm forego the interest their capital would have earned as loan capital. Either the owners of capital earn interest or they receive a combination of dividends and a share in the increasing net worth of the firm. The owners of capital always require payment for the use of their finance during the period of the circuit. However, if there were no economic surplus, resulting from production, these owners would be hard pressed indeed to find ways of achieving this expectation of payment for use of their finance.

Whether interest payments, and still more dividends, should be called *costs* at all depends entirely upon the perspective of the question. From the point of view of the managers of a firm trying to retain as much profit as possible for internally financed growth, they both appear as the costs of having raised external finance in the past, and as the expected costs that would have to be met if more finance were needed in the future. Thus, to the firm they appear as the cost of raising capital. From the perspective of shareholders, on the other hand, dividends are not a cost at all, but their share in profit.

From the perspective of an economist interested in the production and size of the economic surplus, both are clearly claims on the surplus, which is the excess of output over the use of resources and consumption necessary merely to maintain the current level of output. Surplus is the amount which production makes potentially available to be used as investment to increase future output. From the perspective of an economist interested in current economic welfare, on the other hand, neither profits nor wages should properly be treated as a cost, since the objective is to maximise the sum of current net output, that is, value added or income, and these are not costs but incomes.

There is a social cost to using labour, but it is not the wage. Instead, this cost is the disutility of work. Likewise, there is a social cost to using capital, but it is not the dividend or rate of interest. Instead this cost is the cost of replacing capital, fixed capital or stocks of circulating capital, that is consumed as a result of its use. Other things being equal, if production becomes more efficient, both wages per worker and profit per £ of capital can increase – indeed, to allow these increases would be the whole point of trying to raise efficiency.

Another question concerns whether labour and capital are being used as efficiently in their present use as they could be if reallocated to another use. Here what matters is the net output, and therefore income (whether wages or profits), forgone by not reallocating to an alternative use. This is the concept of social opportunity cost. Again, the sum of wages and profits is not a cost to be minimised, but something desirable to be maximised, because wages and profits are simply the result of net output, and its distribution as income. Moreover, whilst productive resources have opportunity costs, money or finance is not itself a productive resource, but merely the power to purchase, and thereby allocate, productive resources.

Performance in the context of growth: the expanding industrial firm

If a firm expands its output each year, it will normally encounter a rising stream of payments for materials, labour and purchases of fixed capital. Its business net cash flow, in our sense, will be significantly less than the economic surplus arising from its production, because these increased purchases of fixed capital will appear as additional expenditures and, hence, deductions from cash flow. The firm's business net cash flow will also be less than its net operating surplus, because in an expanding firm actual expenditure on investment will exceed depreciation provisions.

When industrial capitalists are concerned to increase their profits, it follows from the industrial capital circuit above, that the strategy they adopt is likely to involve cost control. It is generally easier for them to control costs than selling prices, because the economy rarely operates at full capacity. In general there are buyers' markets. Prices are often dependent on competitors. Costs, however, can be controlled internally by improving management techniques, introducing new technology and machines, or limiting wage increases. On the other hand, expansionary strategies will involve increases in certain categories of expenditure (cost), especially investment, marketing, and research and development costs.

Problems may arise, unforeseen or uncontrollable, in the production process, whether of a technical or social origin. The main technical characteristic of the industrial circuit is the process of transformation of the raw materials and inputs into finished goods or services by the application of human work, labour, with the use of buildings, plant and equipment. A further risk taken by industrial capitalists is that the risk of loss of value of their capital increases as the length of the capital cycle increases. This refers to both the period of time taken by the single circulating capital circuit and the extended (multi-circuit) period of circulation of fixed capital. Firms are also exposed to risk due to the uncertainty of the market value of the commodities produced, especially where investment in long-lived fixed capital has been made. Further financial risks and uncertainties are introduced by the use of gearing or debt.

Concluding remarks

Construction firms share the general characteristics of firms engaged in the industrial circuit of capital. However, one relatively special feature of their cash flows derives from the relatively large use of trade credit and trade debt in construction. Payments from clients lag sales by a significant period. Even more significant is the lag between use of some inputs in production and payment from the purchasers (contractors). The effects of these lags dominate the actual cash flows of construction firms.

The final chapter, Chapter 12, therefore turns to the merchant, financial and property capital circuits to review the relationship between capitalist construction firms and their suppliers, clients, banks and landowners.

12
Capital Circuits, Conflicts and Strategies

Introduction

In this last chapter we continue our discussion of capital circuits by describing the capital accumulation of merchant, banking and property capitalists. We look at the different strategies adopted by these different capitalist firms. This approach provides an understanding of the underlying nature of the conflicting interests of the various participants in the production process of the built environment. This is not to say that construction is more prone to conflict than any other sector, though it would appear to be more litigious than many. We argue that construction is more prone to expressions of explicit conflicts of economic interest than many other sectors, precisely because of the complexity of its organisation, its fragmentation and thus the number of interests existing.

We conclude with a discussion of the five different forms of capital engaged in the process of provision of the built environment, which draws together many of the themes of the book.

The merchant (commercial) capital circuit

In the merchant or commercial circuit, illustrated in Figure 12.1, no transformation or production process takes place. In other words there are no changes in the materials or commodities passing through merchants. Consequently their circuit of capital is considerably simpler, and sometimes shorter, than the productive circuit, discussed in the previous chapter. Merchants may be any kind of dealer, retailer or wholesaler, for example, builders' merchants. They use money capital, M, to purchase commodities, C, which they hold in stock until sold to customers at a higher price, M', than was paid originally. The difference, $M'- M$ is known as the merchant's 'turn' and is a gross operating surplus which is then distributed between the merchant firm and the owners of the money capital it uses.

To increase their profits, the major strategies of commercial capitalists are either to increase the margins between the purchasing and selling prices of the commodities handled or to increase the speed with which their capital circulates or rotates, from money back to money. Their annual profit can be thought of as the profit per circuit of their capital multiplied by the average number of times that capital circuit is completed per year. Of course merchants do have handling costs, including the costs of premises, transport and staff, but normally these are low relative to the value of goods traded. Thus control of handling costs per unit of turnover is normally only able to make a secondary contribution to the search for profits. Merchants' profits depend more upon identifying product lines and markets in which margins are widest (the 'Harrods strategy'), or else speed of circulation of capital are greatest (the 'Cash-and-Carry strategy').

The main risks in the commercial capital circuit are the unpredictable price movements in the commodities dealt in by the firm. A risk-averse merchant may hedge these risks by use of futures contracts, whereas risk-seeking merchants may back their judgement of likely price changes against that of the futures market, by buying and selling *on margin*, or by deliberately building-up or running-down stocks of a good. Thus a timber merchant (importer) in the UK who thinks that the price of timber in the UK is going to rise, can deliberately build-up more stocks of timber than the firm expects to sell to their own customers, with a view to re-sale to other merchants if their price forecast is accurate. Alternatively, the merchant could place large advance orders for supply of timber at a future date at a price set today, and which will reflect the general expectation in the market about future price, rather than its own.

A higher proportion of the capital engaged in a pure commercial circuit is likely to be liquid (cash or near cash) than will normally be

Figure 12.1   The merchant (commercial) capital circuit

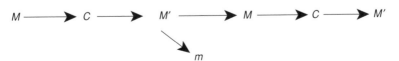

the case for industrial capital. However, if merchants offer trade credit, this will both reduce the liquidity of their capital and lengthen its period of circulation or rotation. Because durable, slowly depreciating fixed capital is not intrinsically involved, commercial capital is more mobile than is industrial capital. There is no necessity for the same commodity be traded in successive circuits, provided the merchant has access to and knowledge of a range of markets.

It has sometimes been suggested that property landlords are engaged in a species of the commercial circuit, buying property at one price and then selling it for a higher price. Whilst it is true that property land-lords do not industrially transform the commodity they purchase, the fact that the commodity is then rented rather than sold introduces a fundamental difference. It is because of the existence of a stock of retained durable capital goods at the end of each circuit, that we treat property capital as a fourth type of circuit, combining some aspects of the commercial and industrial circuits.

The banking (financial) capital circuit

Production usually takes place prior to sale. One important exception to this rule is contract construction. Building work only commences once contracts have been agreed. However, it is usual for suppliers of inputs to be paid before revenue is collected from the sale of the outputs corresponding to those inputs. Sales revenues from this year's output cannot usually be used to pay all of this year's costs of production. If output is expanding, the need for finance other than from pro-ceeds of sales during the period will normally be even greater.

Retained profits from past periods are one source of such finance, but may not be sufficient. Industrial firms therefore find that borrowing finance is necessary to bridge the gap. Credit or cash is needed to pay suppliers and labour before sales revenues have been received. Because of the long lags between acquiring a site, gaining planning permission, the completion of the building process, and the sale of a completed building or property asset, speculative construction requires a greater access to credit facilities than most other industrial activities. Indeed, even when the building is finished, if it is to be rented out, much of the developer's capital remains tied up in the form of real estate. Loan finance is therefore an essential part of the process of providing a built environment by developers or speculative builders. However, in con-tract construction, it is the customer, and not the construction main contractor, who has to find this loan finance.

The third capital circuit is therefore the financial or banking circuit, as shown in Figure 12.2. This is the simplest circuit of all. Money capital, M, is loaned to industrial or merchant capitalists, or indeed to households, either at a stated rate of interest or at a variable market rate. Because interest is charged, the original quantity of money grows to a larger sum, M'. $M' - M$ is again a gross surplus which is shared (after deducting overhead costs) between the banking capitalists, such as the bank's shareholders, and the other owners of money capital, who in this case are the depositors.

Figure 12.2   The banking (financial) capital circuit

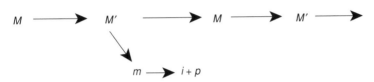

where
  m = interest charged to and received from borrowers
  p = gross operating surplus of the bank from all financial market transactions
  i = interest paid to depositors

A bank mostly uses the money of depositors rather than its own capital. Banks lend money deposited with them by client firms and members of the public. The amount given out in loans is determined by the volume of these deposits.

Banks are essentially money markets in themselves, where those seeking loans are ultimately financed by those depositing funds. Indeed, the accounts of depositors are a bank's liabilities and the loans made by a bank are its assets. The banks thus lend at a higher rate than the interest rate they pay depositors. When the interest rate paid to depositors rises, banking capitalists pass on the higher cost of capital in the form of increased interest rates to borrowers, since banks try to maintain their *turn* or *spread* of interest rates, regardless of the level of the rate of interest. The financial margin is the gap between rates paid on deposits and charged to borrowers.

Not only does a bank use the money of depositors, but it seeks collateral or guarantees for loans in the event of default and it obtains a legally binding contract for the timing and strict repayment of debts. It is the depositors who earn interest on their money, while banks generate profits out of their participation in the process. Banks can only lend once funds have been deposited, though the extent of that lending is in fact greater than the funds deposited. The banking system itself is responsible for creating credit, a point which lies outside the scope of this book.

The exposure to risk of the banking capital cycle is less than in the industrial or commercial circuits. Industrial and merchant capitalists take on the risks of borrowing capital and owning the commodities, which they use to create profits. They accept both financial risk and business risk. A bank, on the other hand, attempts to protect itself from financial risk by matching the terms on which it takes its deposits and makes its loans.

Financial risk for the commercial and industrial circuits refers to the exposure of a net financial position (as borrower or lender) to unforeseen changes in interest rates and the value of money and hence the real cost of debt. In contrast, assuming its depositors retain confidence and its loans remain *good*, a bank is not necessarily vulnerable to the level of interest rates, so long as its financial margin remains intact.

Loans to particular borrowers, or to borrowers in one industry, may become bad debts if business conditions facing that firm or industry deteriorate. In this sense the bank, too, is exposed to business risk. Banks normally seek to protect themselves from this kind of risk by limiting the gearing ratios of their borrowers, and requiring high ratios of interest cover. They also take security for their loans in the form of claims on assets of the borrower whose value it is supposed will survive the business failure of the borrower. However, in the event of a general failure of businesses across the board, as a result of an economic slump, banks are still vulnerable because these real assets held as security may lose their value. This was the cause, for example, of the wave of bank collapses in the USA in the early 1930s.

It is the lending function of banks which controls the direction of an economy, since banks only extend credit to those enterprises and projects of which they approve. Morishima (1984) defends the role and strategic economic power of banks insofar as they do act as investigators of investment plans. He argues that money and resources would be wasted if banks were to lend indiscriminately to inefficient firms or for loss making schemes. In fact, recent criticisms of banks tend to argue that they have been too uncritical in their lending. They have been too keen to lend to borrowers prepared to pay some extra points in interest, and therefore some banks have been prone to understate the risks involved and relax their scrutiny of borrowers' investment plans. This has been the case, especially, in lending to property developers during property booms. Banks have frequently placed too much emphasis upon the supposed safety net provided by certain kinds of collateral, and too little on the intrinsic profit prospects of an investment project. Further criticism of the banking sector has been based on their short term view of investment proposals, favouring projects offering quick, and therefore supposedly more predictable, returns over those offering less predictable but potentially vast economic returns, such as investments in developing new technologies.

For maximum profits, the major strategy of financial capital or banks is to borrow and lend as much as possible.

The landowning (property) capital circuit

In the property capital circuit in Figure 12.3, money, M, is used to purchase property, $C(RE)$. Once purchased the property assets are then rented out in return for a net annual rental income, n, but ownership of the asset is retained by the property capitalist. This circuit of capital may continue for several years until the owner decides to sell the property, usually for a larger amount, M', than it cost to purchase.

The main characteristic of the property capital circuit is that the ownership of the real estate, RE, remains with the property owner and income is derived from renting the property in return for its use by tenants, as well as any speculative gain, which is realised when the property is sold.

Figure 12.3   The property (real estate) capital circuit

$$M \longrightarrow C(RE) \longrightarrow n + C(RE) \longrightarrow \ldots \longrightarrow n + C(RE) \longrightarrow M' \longrightarrow M$$
$$\searrow$$
$$m$$

where
$\qquad C(RE)$ = commodity real estate
and
$\qquad n$ = net rental income

Because there are two sources of income, rents and income from resale, occurring in separate periods, the return derived from rents is usually lower than the opportunity cost of capital, depending on the expected discounted gain from the eventual sale of the property. The strategy of property capitalists is therefore to time the sale of assets to attempt to maximise the speculative gain. To enable the property owner to obtain the best perceived timing of the sale, rent income is used to cover intervening net costs of interest payments and maintenance.

All the time the building is owned by the property owner debt is mobilised to obtain greater returns on the firm's own capital invested. This is achieved by relatively high levels of gearing, based on asset values and the ability of the property owner to service debt out of income. Not all of a property's valuation can be used as collateral for funding purposes. Otherwise, lenders would be exposed to the risk of bad debts, if property owners were to encounter cash flow difficulties and the properties were to be sold below their valuations in a depressed market. The amount of finance available to property capitalists therefore depends on their existing financial position and the expected state of the property market.

Difference of circuits and conflict between firms

One significance of this approach using the different capital circuits of industrialists, merchants, banks and property owners is that while they are all capitalists, that is, they all require capital and then use it to accumulate more capital, they are nevertheless in fundamental conflict with each other. One conflict occurs in the market between merchants and producers. Like buyers and sellers in any market, capitalist producers want to buy their material inputs at low prices, while it is in the interests of their suppliers to sell their stock at high prices. Similarly, when manufacturers wish to sell their finished goods to merchants, the roles of buyer and seller are reversed. This time it is the producer who wants a high price, while the merchant attempts to achieve a low one.

Another conflict arises between producers and their banks, especially if producers are, on balance, net borrowers. The greater the interest rate, the higher the proportion of the economic surplus created by producers that goes to banks, leaving a smaller share of the surplus for the producer. Similarly, the higher the rent, the less of gross profit is available as net profit to the tenant industrialist, merchant or bank. As most firms prefer to use their financial resources to exploit the opportunities in their own particular markets, it makes sense for them to rent their premises and so obtain a higher return on their capital by investing in their own areas of technical expertise. Nevertheless, renting brings tenants into conflict with their landlords. This sense of conflict between capitalist firms is distinct from that in which firms compete with each other within markets. For instance, contractors compete with each other, as do brick manufacturers, building merchants and banks.

A complex unity of circuits: the modern construction corporation

So far, this approach to capital circuits has distinguished between different types of firms, but it can also be applied within firms. The typical large construction firm is a complex combination of industrial, commercial, financial and property circuits.

Insofar as the firm directly controls and engages in production, even if this is only a small part of the totality of the built environment production process, then part of the firm's capital follows the industrial circuit. This is the case for a firm undertaking design services or construction management services as well as for one directly employing operative construction workers. All that is required is that workers of some kind be employed to produce an output which is a transformation of the inputs used in its production and which has a market as a commodity, and therefore a market price. The commodity output does not, of course, have to be a recognisable final product from the point of view of a building owner.

Insofar as the firm operates as a commercial intermediary, however, buying outputs from other firms, adding its mark-up and reselling them in another market for a higher price, such that the functions of its own staff are to manage the *transactions* involved, then the firm is operating as a kind of merchant, and part of its capital follows the commercial circuit. This might apply, for example, to a firm which obtains main contracts from clients which it fulfills entirely by splitting them up and then buying-in design or construction services from subcontractors. Ball (1988) raised the question of whether modern main contractors should be regarded as merchants rather than producers. Our answer is that it all depends on whether the firm in question actually undertakes production of some kind, even if only the production of a work plan for work that will be executed by others, or whether it merely writes and places a set of contracts with those others, in which those contracts specify the output to be produced, but leave the organisation of that production to the subcontractors. In practice, on any given project a management contractor will normally be doing a limited amount of production and a lot of transaction management, but these proportions may vary from firm to firm and project to project.

That most large construction firms are engaged in the property circuit is self-evident. A glance at their accounts reveals a rough estimate of just how high a proportion of their capital is engaged in this circuit, in the ownership of real estate held as an investment.

Finally, what of the financial circuit? This, it might be thought, is quite outside of the scope of the activities of a construction firm. However, almost all large firms possess some financial assets in the form of loans made outside of their own business. Moreover, the relation of the corporate headquarters to its operating divisions is often one of a banker to borrowers, whilst its relation to its shareholders and banks (and, sometimes, even to its cash-generating subsidiaries) resembles that of a bank to its depositors.

According to Goold and Campbell (1987), the corporate centre is essentially a relatively specialised financial institution. It is specialised in the sense that it conducts financial business mostly with a narrow class of borrowers. It therefore possesses more information about them than could a general lender. These borrowers are its own subsidiaries or operating businesses. Requiring finance for an investment, they are required to bring their proposals to the corporate centre, which has a monopoly of the banking role as far as they are concerned. The centre allocates funds. In effect it makes loans to those proposals which meet its investment criteria as a bank. These criteria will differ perhaps from those of a high street bank, being rather more like a supplier of venture capital, such as 3i.

Capital supplied to an operating business is expected to pay a high target rate of return. This return will then have to be paid to the centre, in the form of an increased central appropriation of the business's operating profit. If the business manages to make even more profit than this target sum from its investment, its managers may or may not be allowed to retain control of some of it, as an incentive. If the business falls short of the profit target set on a particular investment, it may make up the deficit, if possible, from surpluses over target on other activities. If it cannot do this, its 'bank' may move in to impose new management.

At the same time, outsiders lending the diversified corporation finance, or buying its bond issues, do not know in which particular business their money is to be used, still less the particular investment project. Instead they are, as it were, lending money to an institution which *guarantees* them a certain return, and is able to do so because of its size and diversity. The corporation can thrive so long as it can systematically attract 'deposits' in this way at a lower cost of capital than the return it can extract on its 'loans' to its operating businesses. This difference becomes profit attributable to its shareholders, and explains how the return on shareholders funds can be higher than the average rate of return on operating assets made by its subsidiaries.

Ive (1994) shows how large have been the flows of funds, between subsidiary and centre and centre and subsidiary, *within* large construction groups, and the extent to which they do act as financial institutions in the way described. Business ratio reports (ICC; Jordans; Dun & Bradstreet) show the high levels of external debt as a ratio of shareholders' funds that characterised the large construction groups in the late 1980s and early 1990s. Goold and Campbell (1987) show how the financial control style worked in two construction sector corporations in the 1980s, Hanson and Tarmac.

Every multi-circuit firm is a combination of three kinds of management. Each of these may usefully be thought of as a bureaucratic department, with its own function, objectives and modes of measuring performance. *Production management* seeks to control costs by relating them to a notional value of output. *Commercial management* attempts to widen margins on turnover without reducing internal production costs, either by finding higher prices for outputs or cheaper prices for inputs; or to speed the rotation time of the industrial and commercial capital circuits. These together influence the rate of return on operating capital. *Financial management* attempts to widen the ratio between the return on shareholders' funds and the return on operating capital.

It is always possible to break any *return on capital*-type ratio into two components, one measuring *margin* in the form of profit per unit of sales, and the other measuring speed of rotation, in the form of sales value obtained from use of a given amount of capital. One of the most famous versions of this is known as the DuPont Formula. Profits and assets can be measured in a range of ways. So long as the concept of profit used corresponds to the measure of assets used, any formula of this type remains valuable as a source of insight into a firm. By definition,

$$\pi/K = \pi/T \,.\, T/K \qquad\qquad (12.1)$$

where
> T = turnover or sales
> K = capital or assets

and

> π = profit.

For present purposes, let us define K as the net assets used by the firm, rather than either gross assets or shareholders' funds, and π as operating profit before interest. π/K can then be called the return on net assets or RONA. The relation of the return on shareholders funds (ROSF) to RONA is then clearly dependent on the ratio of the cost of borrowing to RONA, and on the gearing ratio:

$$ROSF = (\pi - I)/SF \tag{12.2}$$

where
$$I = \text{interest bill}$$
and
$$SF = \text{shareholders funds}$$
but
$$I = i.B \tag{12.3}$$
where
$$i = \text{interest rate}$$
and
$$B = \text{value of borrowing or debt}$$
and
$$\pi = r.K \tag{12.4}$$
where r = the rate of return on the assets
Now

$$SF = (K - B) \tag{12.5}$$

Substituting (12.5) in (12.2),

$$ROSF = (\pi - I)/(K - B) = [(r.K) - (i.B)]/SF \tag{12.6}$$

and $\quad ROSF = r + \left[(r - i).\dfrac{B}{(K - B)}\right]$

$$\tag{12.7}$$

Financial management attempts to widen the ratio between the rate of financial return on shareholders' funds and the rate of return on operating assets, by raising the gearing ratio, (B/K) while controlling the cost of capital to the firm (I/π).

The calculations within each of these management departments are also distinct. In production management the mode of calculation is related to average prices and costs and the quantities and prices of resources used, such as the cost and the number of machines. This mode of calculation is relatively insensitive to time. That is, whilst a time discount rate may be imposed (perhaps by the financial management) in investment appraisal, the mode of calculation is otherwise indifferent to matters of time profile. Investments will be sought if they seem likely to widen the long run gap between average cost and average selling price. This might be achieved by investments which reduce unit costs, even if they have the disadvantages of tying up capital and increasing the total use of funds. There may be a capital-using bias to technical progress, such that cost saving proposals from production management more often than not tend to raise the ratio of capital employed to turnover. Long run optimum methods of production may tend over time to become increasingly roundabout, such that they require relatively more fixed capital. Fixed investment pay-back periods may also lengthen, so that more capital is tied up for longer. Production initiatives will tend to identify opportunities for resource saving process innovation, rather than product innovation. Unit prices of inputs will tend to be taken as given, because they are outside the influence of production management. Consequently the emphasis will be on reducing input quantities used per unit of output.

Commercial management includes both marketing and procurement. In terms of marketing, the calculations are based on day to day prices in market places and on the contract price of goods and services. The cost of time is recognised as the cost of holding stocks. Marketing initiatives tend to identify opportunities to sell new products or find new markets for existing products. The common feature of these new products or markets is that they offer a higher than current selling price per unit of production cost. However, these marketing initiatives may share something of the impact of production initiatives on financial requirements. They may require increased current expenditures, for the prospect of higher margins in the future. It is a costly exercise to invest in entering new markets, establishing customer awareness of a firm, building up market share, and other marketing objectives.

In terms of procurement, initiatives tend to identify possibilities for input source substitution. Commercial managers often replace input suppliers, whose relative price has risen, or quality fallen, with cheaper or higher quality alternative sources for those inputs. The typical calculation is simply either in terms of price minimisation or value for money. Value for money measures quality relative to cost in terms of each pound spent on purchasing.

Finally, in financial management the problem is to establish the amount of capital required, the sources of funding and the most profitable application of the funds available. It is the function of financial management to find both the minimum amount of capital required to operate the business and the cheapest way of financing it. The starting point of financial management is therefore not costs and prices but cash flows and thus the amount of capital required. Financial management initiatives tend to seek ways of speeding the rotation of the firm's operating capital, and thus reduce capital requirements relative to turnover. This may even be at some cost in terms of higher unit production costs or lower commercial margins.

Note, in passing, that there are two kinds of initiative which are not mentioned above, as the typical initiatives of any *one* management department acting alone. These are product innovation, and neo-classical-type factor-substitution, as opposed to input-*source* substitution. Product innovation typically requires joint initiative by marketing and production management, and neo-classical-type factor-substitution requires joint initiatives by procurement and production management.

The forms of capital

This book, like its companion, *The Economics of the Modern Construction Sector*, has been concerned with the behaviour of firms involved in all aspects of the production of the built environment. Central to our understanding of their activities is their management and accumulation of capital. We are now in a position to combine and summarise the different aspects of the construction process and the operation of individual firms within that process, by focusing on capital. Capital may take five different forms. They are:

- money
- shares and bonds (financial assets)
- commodities
- productive resources
- commodities in the course of production.

The history of a firm begins with capital in money form. Money is the *first* of the five forms taken by capital. Fiduciary paper monies (bank notes) are of course themselves only a kind of paper. Once upon a time, analogously to other financial assets, paper money gave its holder a claim on commodities or money owned by the issuer – in this case *commodity-money* (gold or silver) held by the issuing bank. This, however, is no longer the case. Within the boundaries of a nation state, that state's bank's notes are legally universally acceptable means of payment of debts and of exchange for commodities. They represent capital in its most liquid, least specific, most abstract and general form. All capital at some points in its life of continuous mutation, takes the form of money. The particular money in question here has been exchanged for ownership of shares in a newly born firm. These shares have been created from nothing, simply on the strength of a company prospectus. This statement of intent describes how the firm plans to use the money obtained from the sale of shares, to make more money, which in turn would then belong to the share subscribers.

Shares and bonds are the key instances of capital in its *second* form. As a *paper financial asset* a share or bond gives its holder legal title to a share in the money income or real assets of the entity which issued the paper. In the event of default, the owner of a paper financial asset would have a legal claim on the commodities owned by the issuer. For the issuer the paper is a financial liability.

How does a firm, which issues financial liabilities, obtain the stream of money income from which it can meet these claims? It does this in one of three ways.

1. The first way is by exchanging the money it now possesses for commodities, and then, somehow, re-exchanging those same commodities for a larger sum of money. This is the mode of the merchant. Capital following this sequence of forms is called merchant or commercial capital. The *commodity-form* is the *third* form taken by capital.

2. The second way is by lending the money, at interest, to a borrower. To secure their loan and create a legally enforceable claim on the income or assets of the borrower, the act of lending usually involves exchange of money for a paper liability issued by the borrower, and held as a financial asset by the lender until the loan is fully repaid. If a firm raises money by share issue and then uses that money to make loans, then the rate of interest collected on those loans, less the operating expenses of the firm, sets the rate of profit on its share capital. This is the mode of the banker. Capital following this sequence of forms is called banking or financial capital.

3. The third way is by using the money to buy a combination of commodities and productive services. These then need to be supervised during the *productive transformation* of these commodities as a necessary aspect of the process of production of new commodities. For example, the productive process involves the transformation of the tangible form of commodities from, say, clay to bricks or bricks to houses. Finally, the new commodities need to be sold for money or

financial assets. The profit, and hence the ability to pay a money return to shareholders, comes from the excess of the money value of the commodities produced and sold over the money value of the commodities and productive services purchased and used up. This is the mode of the industrialist. Capital following this sequence of forms is called industrial capital.

There is a fundamental distinction between a commodity and a productive service. A commodity consists of tangible, durable (and therefore re-sellable) goods of some kind. A productive service is the intangible, unstorable, evanescent potential output of productive resources, of which by far the most important are the services of labour. The services of productive resources are the capability they have to be used to produce other commodities. Once the services of productive resources have been purchased, they cannot be re-sold as such. The capitalist who exchanges money for productive services, therefore, whether in the form of labour or the hire of a piece of machinery, for the moment has neither money nor commodities. The capitalist who exchanges money for goods, on the other hand, still has their capital in a tangible and exchangeable form, the form of a commodity. There is, therefore, a *fourth form* taken by capital, the form of *productive-power* or *productive-resources*.

As commodities are used-up and transformed in production, at the very moment they disappear in their former commodity-form they may reappear in a new commodity-form, or there may be an elapse of time (for example between the opening of a bag of cement and the appearance of a finished, sellable concrete floor). To allow for this we need a *fifth form* of capital, namely *commodities in course of production*. Capital in this form appears in the accounts of industrial firms as part of the wider item of the 'value of work in progress'.

Note that it is only in the mode of industrial capital that capital passes through all five forms. The term for a process of trans-formation of capital, its passage from form to form, is a *circuit of capital*. The term circuit is apposite because it stresses the circularity of a process that ends with capital in the same form in which it began the process – the money form. The term is too well-established to change now, but even more apposite would be *spiral* of capital – to stress both the circularity and the expansionary aspects of the process over time.

If we take a still photograph, as it were, of capital in this process of circulation at any one moment, the capital of a particular industrial firm will comprise parts in each of the five forms. This is indeed shown when a balance sheet is drawn up.

Now, it becomes clear that the significance of the distinction between capital forms is two-fold. On the one hand, it enables us to identify and distinguish the four basic circuits of capital as the four different ways in which capital can grow in value. This, of course, is fundamental to a typology of possible capital strategies. These are the strategies available to the owners of capital who wish to accumulate wealth. On the other hand, the distinction between forms also enables us to see the sources of the uncertainties and the limits to the opportunities faced by the controllers of a particular set of types of capital.

Let us consider the five forms of capital in turn:

(1) *Capital in money form* is liquid. It is in principle free to flow from one venture or specific activity to another under the direction of its owners. Moreover, money is also abstract or without a definite concrete shape. It is *abstract value* with the as yet unrealised potential to be converted not only into any of the other forms of capital, by a simple act of exchange, but also to be applied to any concrete line of business, any specific commodity or production process.

The potential return to capital in money form does not, therefore, depend upon the continued profitability of the activity out of which it was generated. In practice, however, the industrial firm acts as a constraint on that potential mobility, because the controllers of such firms plough back most of the money capital resulting from successful completion of the circuit of industrial capital into the continuation, preferably on an expanded scale, of the same specific activity. Certainly it is true that over a period of time, most firms diversify and change their activities. But the pace of such change is almost glacial by comparison with the potential mobility inherent in the money form.

The risks to the *value* of capital in money form come only from general threats to the security of national paper money as a store and measure of value. Inflation is simply a general reduction in the national purchasing power, in terms of commodities or productive services, of national money. Other risks to the value of money capital come from the loss of convertibility or currency devaluation, which lead to a reduction in the international purchasing power of national money.

(2) *Capital in financial asset form* is somewhat less liquid than money capital. Whilst there are markets in ownership rights over existing

financial assets, such as the stock market, asset prices in these markets are prone to considerable short term fluctuations. The range in value from the minimum to the maximum for a financial asset is usually considerable. The minimum value the owners are confident they can obtain in the worst scenario, which they can foresee as plausible, is very different from the maximum value they may realistically hope to obtain if the timing of its sale were to coincide with a peak in its market price. Moreover, the financial liabilities of a particular issuer only retain any value so long as that issuer remains solvent or at least remains the owner of realisable assets. Thus, the valuation put on a financial asset is based on expectation, and is subject to uncertainty.

(3) *Capital in finished commodity form* is semi-liquid. It needs only an act of exchange, for which it is already in a suitable form, to be converted into liquid, money form. Its value is measured by anticipating the expected sum of money for which it will be exchanged. The risks to this valuation come from both general and specific threats to the exchangeability of commodities for money. There may be a general glut or excess of all commodities on the market relative to monetary effective demand. Prices of commodities may or may not fall in this situation. Even if they do fall, the glut may continue. If prices do not fall in such a situation, the value of commodities must still be recognised to have fallen, in that *value*, for a commodity, comprises both a price and marketability at that price. Even if prices are *sticky*, in such a glut situation, stocks of commodities lose much of their potential liquidity or realizability in money form. This will most likely be reflected in the downward revaluation of such stocks in the accounts of their owners. This problem is the Keynesian one of lack of effective demand. Alternatively, there may be a specific lack of demand relative to supply for the particular commodity in question, either because of the emergence of substitute commodities or an over optimistic or unplanned increase in supply. If the commodity in question is not a final consumption good, the lack of demand may derive from a lack of demand for, or over-supply of, another commodity. Risks to the value of capital in commodity form here come chiefly from the production or supply of direct competitors and from the suppliers of substitutes.

Contractors do not, by definition, face the main risk of being unable to sell finished output, precisely because they already have a contract of sale before they commence production. The risk that the commodity is no longer wanted, or has fallen in value since the contract was signed, falls entirely upon the customer. Though contractors' output is pre-sold, it is not, however, pre-paid. If a fall in value-in-use or resale-value is great, and if the construction project's contract value is high relative to the total assets of the customer, the contractor still faces a residual risk. Their customer may be unable to pay, and may have insufficient net assets for the debt to be effectively retrievable through the courts. This makes work for property developers the most vulnerable to becoming a bad debt for reasons deriving from a general reduction in the value of buildings. This risk is more serious for construction firms than the relatively diffuse risk of customer bankruptcy and consequent non-payment inherent in any individual business transaction, because it is liable to affect many of their customers simultaneously.

(4) *Capital in the form of productive resources* faces yet other kinds of risk and uncertainty affecting its valuation. Its value is the value of the output to be obtained from the productive resources over the period for which they have been hired or bought. This value depends on the rate of utilisation of resources, the amount of work actually done compared to the potential use of resources. Their capital value also depends on the efficiency of each unit of that work, which is the value of output achieved per unit of work done. For example, workers may be hired for an 8-hour working day, but the firm may only find and organise productive work for them to do in 6 of those hours. The value of output produced in each of those hours may then itself vary, either because of variation in physical productivity (units of output per hour), or because of variation in the value of each unit of output.

Utilisation rates to some extent depend upon social convention and regulation. Night-working on sites in residential areas may not be permitted. Labour market institutions may require employment to be for a fixed length of working day. However, utilisation depends also upon the organisational efficiency of the firm in achieving as close as possible a correspondence between resources available to work and tasks ready to be done. Once this is achieved, the value of output per worker productive hour or per crane productive hour will also depend upon the market conditions under which the output is sold. This determines the value of a unit of output. The efficiency with which the production process has been designed, influences the number of units of output produced per hour. In the case of the worker, but not of the crane, output per hour will also depend upon another variable, the intensity with which the worker works, or the amount of effort expended per hour. Work intensity and motivation are, in turn, subject to influence by factors specific to the organisation.

The key point is that it is by no means the case that certain inputs, once hired, automatically or always produce the same output per period of time, although this is sometimes implied by economists who construct models in terms of production functions. In no industry, perhaps, is this less the case than in construction. In the worst situation, production may come to a complete standstill, despite resources having been hired, perhaps because of a major accident, or a strike, or a failure in the supply chain preventing necessary inputs from arriving. More normally, there will be a production plan, specifying the output expected per period, but the management of production is a continual series of responses to correct, if possible, shortfalls of actual output below planned levels, arising from a multitude of causes.

(5) *Capital in the form of unfinished output* has its own problems of uncertainty of valuation. Not yet a commodity, it is not liquid in the sense of being realisable as money by sale to a customer. Nor can it be withdrawn from its intended use and be redeployed. The balance sheets of many construction firms are dominated by capital in this form. The contracting system has developed its own practices for its conventional valuation. The chief of these is the Bill of Quantities, used to give an interim valuation to output of unfinished buildings. Many of the particular complexities of construction firms' accounts derive from the fact that output on sites is valued by reference to the Bill of Quantities every month. Certificates for interim payment related to this valuation are issued, and this value then appears in the firm's accounts as 'value of work done', for which payment is due. Up to this point, unfinished output appears as 'stocks and work in progress'. Once valued, however, it appears as 'trade debtors'. Nevertheless, prudence may dictate that not all the expected value of output can be treated as acknowledged debt. Consequently, a provision may be made in the firm's accounts, to allow for the fact that the final valuation of all the work done on a project has not yet been agreed, and there is still scope for dispute between the contractor's and client's views of what that value will be.

Such discrepancies of valuation come to the surface, mainly, at the point contractors make their last claim for payment. This last claim for payment is the balance between the sum of interim payments already made and their view of the total amount they should receive, including monies deducted or 'retained' by clients from those interim valuations. If finally, after dispute and negotiation, they settle for a lesser sum than they claimed, the entire difference between expected and out-turn value of output on a project falls upon that last time period's accounts. The shortfall in turnover becomes a reduction in the value of net current assets, unless a 'provision' has already been made in current liabilities.

In the accounts of construction firms, turnover normally relates to the amount contractually receivable from customers for the value of work done by the firm over the period. It does not therefore measure cash received from customers over the period. Moreover, some of the valuations of the work done will be those of the firm itself, and not yet accepted by the customer.

Yet, the system of sale-before-production that characterises contracting, together with the practice of interim payments, leaves the contractor in a very different and more favourable position, as regards the valuation of unfinished output, compared to a speculative builder. For the speculative house builder, for example, the value of work done including completed dwellings, only becomes *turnover* as and when customers for those dwellings are found and contracts to purchase are completed.

Concluding remarks

Using ideas of the different kinds of capital circuits, and the different risks to which capital in different forms is subject, it is possible to bring together or synthesise the (otherwise disparate) range of factors that affect the accumulation of capital in firms engaged in the construction process.

This book has attempted to focus upon the behaviour of, and strategies open to, construction firms engaged in the industrial circuit of capital. However, in order to do this, it has been necessary to place industrial construction firms in their broader context, in particular the context constituted by firms engaged in the financial and property circuits of capital.

Property rents and interest on construction and property loans each dwarf the total net profits of industrial firms engaged in construction (firms in the construction, building materials, components and design industries). Measured this way, industrial capital is the smaller part of the total capital obtaining returns in the construction and property sector. Yet in our view, because of its role in shaping the production process, that industrial capital is potentially the most important and influential, both in shaping what is and isn't built, and in shaping how it is built, and with what consequences for the millions of people who earn their livings in this process of production of the built environment.

But the significance of the other circuits of capital for the construction process does not end with their sharing in the economic surplus of this sector. In addition, they may offer relatively alluring alternatives compared to the difficult and uncertain paths open to industrial capital to follow in large parts of the construction sector.

There is, if we may put it like this, an ongoing temptation to construction sector firms to seek to transform themselves into something else: into merchants ('merchant contractors'); into finance houses ('financial holding companies'); into property companies, rather than to surrender liquidity for their capital by making irreversible, sunk-cost-type production investments and expenditures.

But it is not only in the world of financial assets that a choice must be made between the attractions of liquidity and those of rate of return. Construction firms too can choose to seek the path of greater liquidity – for example by speeding the rotation time of their capital, reducing their investments in specific tangible and intangible assets, and concentrating on methods of operation which bring the moment of payment (cash inflow) forward in time as much as possible relative to the moment of production and cash outflow.

But the price to be paid for the firms is the alternative forgone – that is the higher margins (from lower production costs and/or higher selling prices) that result from successful future-orientated expenditures and investments in competitive advantage. For the industry's customers, the price paid may show itself in a relative shortage of industry capacity in periods of expansion of demand, and in a slow rate of improvement in productivity and efficiency – matters that will also show themselves as problems for the sector's workforce, which instead may face a trade-off, in increasingly unfavourable conditions, between level of earnings, work intensity and job security.

Bibliography

Abramovitz, M. (1968). 'The passing of the Kuznets cycle', *Economica*, 35, pp. 349–67.

Akintoye, A. and Skitmore, M. (1991). 'Profitability of UK construction contractors', *Construction Management and Economics*, 9, pp. 311–25.

Alchian, A. and Demsetz, H. (1972). 'Production, information costs, and economic organization', *American Economic Review*, 62, pp. 777–95.

Anderson and Gatignon (1986). 'Modes of foreign entry: a transaction cost analysis and propositions', *Journal of International Business Studies*: Fall.

Andrews, P.W.S. (1964). *On Competition in Economic Theory*, London: Macmillan.

Andrews, P.W.S. and Brunner, E. (1975) *Studies in Pricing*, London: Macmillan.

Arestis, P. and Kitromilides, Y. (eds) (1989). *Theory and Policy in Political Economy: Essays in Pricing, Distribution and Growth*, Aldershot: Edward Elgar.

Arestis, P. (1992). *The Post-Keynesian Approach to Economics*, Aldershot: Edward Elgar.

Arestis, P. and Chick, V. (eds) (1992). *Recent Developments in post-Keynesian Economics*, Aldershot: Edward Elgar.

Arestis, P. Chick, V. and Dow, S.C. (eds) (1992). *On Money, Method and Keynes*, London: Macmillan.

Armstrong, P., Glynn, A. and Harrison, J. (1991). *Capitalism since 1945*, Oxford: Blackwell.

Bails, D.G. and Peppers, L.C. (1993). *Business Fluctuations: Forecasting Techniques and Applications*, 2nd edn, Englewood Cliffs: Prentice-Hall.

Ball, M. (1983). *Housing Policy and Economic Power: The Political Economy of Owner-occupation*, London: Methuen.

Ball, M. (1985). 'Coming to terms with owner occupation', *Capital and Class*, 24, pp. 15–44.

Ball, M. (1988) *Rebuilding Construction*, London: Routledge.

Ball, M. (1994). 'The 1980s property boom', *Environment and Planning A*, 26, pp. 671–95.

Ball, M. and Wood, A. (1994A). *Trend Growth in post-1850 British Economic History: The Kalman filter and historical judgement*, South Bank University, Working Paper.

Ball, M. and Wood, A. (1994B). *Housing Investment: Long-run International Trends and Volatility*, Birkbeck College Discussion Papers in Economics, 11/94.

Ball, M., Wood, A. and Morrison, T. (1994). 'Structures, Investment and Economic growth: a long-run international comparison,' South Bank University, Working Paper.

Ball M. *et al.* (1998) *Economics of Commercial Property Markets*, London: Routledge.

Banwell, H. (1964) *The Placing and Management of Construction Projects*, Report of the Committee on the Placing of Management Contracts for Building and Civil Engineering (the Banwell Report), London: HMSO.

Barlow, J. *et al.* (1997). *Towards Positive Partnering*, Bristol: Policy Press.

Barrett S.M. *et al.* (1990) 'The land market and the development process: a review of research and policy', *Occasional Paper*, 2, School for Advanced Urban Studies, University of Bristol.

Baumol, W.J. (1982). 'Contestable markets: an uprising in the theory of industrial structure', *American Economic Review*, 72, pp. 1–15.

Beer, S. (1972). *The Brain of the Firm*, London: Allen Lane.

Builders Economics Research Unit (1974). *The Building Timetable: the Significance of Duration*, London: UCL.

Bhaduri, A. (1986). *Macroeconomics: The Dynamics of Commodity Production*, London: Macmillan.

Bishop, D. (1975). 'Productivity in the construction industry', in Turin, D. (ed.) *Aspects of the Economics of Construction*, London: Godwin.

Bon, R. (1989). *Building as an Economic Process*, London: Routledge.

Bon, R. and Crosthwaite, D. (2000). *The Future of International Construction*, London: Telford.

Bootle, R. (1996). The Death of Inflation, London: Nicholas Brealey.

Bowles, S. and Edwards, R. (1993). *Understanding Capitalism*, New York: Harper-Collins.

Bowles, S., Gordon, D. and Weisskopf, T. (1989). 'Business ascendancy and economic impasse', *Journal of Economic Perspectives*, 3, pp. 107–34.

Bowley, M. (1966). *The British Building Industry: Four Studies in Response and Resistance to Change,*Cambridge: Cambridge University Press.

Braverman, H. (1974). *Labor and Monopoly Capital*, New York: Monthly Review Press.

Brealey, R.A. and Myers, S.C. (1996). *Principles of Corporate Finance*, 5th edn, New York: McGraw-Hill.

Brenner, M. (1998). 'The economics of global turbulence', *New Left Review*, 229, pp. 1–265.

Buzzell, R.D. and Gale, B.T. (1987). *The PIMS Principles: Linking Strategy to Performance*, New York: Free Press.

Campagnac, E., Lin, Y.J. and Winch, G.M. (2000). 'Economic performance and national business systems – France and the UK in the international construction sector', in Quack, S., Morgan, G. and Whitley, R. (eds), *National Capitalisms: Global Competition and Economic Performance*, Dordecht: Benjamin.

Canova, F. (1993). 'Detrending and business cycle facts', *CEPR Discussion Paper*, London.

Central Statistical Office (1993). *UK National Accounts, Sources and Methods*, (4th edn), London: HMSO.

Central Statistical Office (1993). *Report on the Census of Production 1991, PA 500: Construction Industry*, London: HMSO.

Chamberlin, E.H. (1933). *Theory of Monopolistic Competition*, Cambridge, Mass: Harvard University Press.

Chick, V. (1983) *Macroeconomics after Keynes*, Cambridge, Mass: MIT Press.

Coase, R. (1937). 'The nature of the firm', *Economica*, 4, pp. 386–405.

Coase, R. (1960). 'The problem of social cost', *Journal of Law and Economics*, 3, pp. 1–44.

Cole, K., Cameron, J. and Edwards, G. (1991). *Why Economists Disagree*, London: Longman.

Construction Industry Group (1981). *Construction Industry into the 90s*, London: Institute of Marketing.

Coombs, R., Saviotti, P. and Walsh,V. (1987). *Economics and Technical Change,* London: Macmillan.

Cowling, K. (1982). *Monopoly Capitalism,* London: Macmillan.

Cullen, A. (1983). 'The changing interface between the construction industry and the building materials industry in Britain', *Production of the Built Environment, Proceedings of the Bartlett International Summer School,* 4, pp. 3.45–3.53, London: UCL.

Cyert, R.M. and March, J.G. (1970). *A Behavioural Theory of the Firm,* Englewood Cliffs: Prentice-Hall.

Dahlman, C.J. (1979). 'The problem of externality', *Journal of Law and Economics,* 22, pp. 141–62.

Darwin, C. (1929 edn). *The Origin of Species by Means of Natural Selection,* London: Watts.

Department of Employment (Quarterly to 1995). *Employment Gazette,* London: HMSO.

Department of Environment Transport and the Regions (formerly Department of the Environment) (annual) *Housing and Construction Statistics,* London: HMSO.

Department of Environment, Transport and the Regions (formerly Department of the Environment) (monthly) *Monthly Statistics of Building Materials and Components,* London: DETR.

Devine, P. *et al.* (1985). *Introduction to Industrial Economics,* 4th edn, London: Unwin Hyman.

Dietrich, M. (1994). *Transaction Cost Economics and Beyond: Towards a New Economics of the Firm,* London: Routledge.

Drewer, S. (1980) 'Construction and development: a new perspective', *Habitat International,* 5.

Druker, J. and White, G. (1995). 'Misunderstood and undervalued? Personnel management in construction', *Human Resource Management Journal,* 3, 3, pp. 77–9.

Drucker, P.F. (1994). *Managing for Results: Economic Tasks and Risk-taking Decisions,* Oxford: Butterworth-Heinemann.

Drucker, P.F. (1995). *Managing in a Time of Great Change,* Oxford: Butterworth-Heinemann.

Earl, P.E. (1983). *The Economic Imagination: Towards a Behavioural Analysis of Choice,* Brighton: Wheatsheaf.

Earl, P.E. (1986). *Lifestyle Economics: Consumer Behaviour in a Turbulent World,* Brighton: Wheatsheaf.

Egan, J. (1998). *Rethinking Construction,* London: DETR.

Eichner, A. (1979). *A Guide to Post-Keynesian Economics,* London: Macmillan.

Eichner, A. (1980). *The Megacorp and Oligopoly: Micro Foundations of Macro Dynamics,* New York: M.E. Sharp.

Felstead, A. (1993). *The Corporate Paradox: Power and Control in the Business Franchise,* London: Routledge.

Fine, B. (1975). 'Tendering strategy', in Turin, D. (ed.) *Aspects of the Economics of Construction,* London: Godwin.

Fine, B. and Leopold, E. (1993). *The World of Consumption,* London: Routledge.

Goold, M. and Campbell, A. (1987). *Strategies and Styles: The Role of the Centre in Managing Diversified Corporations,* Oxford: Blackwell.

Gordon, D.M. *et al.* (1994). 'Long swings and stages of capitalism', in Kotz, D.M. *et al.* (eds), *Social Structures of Accumulation,* Cambridge: CUP.

Gore, T. and Nicholson, D. (1991). 'Models of the land development process: a critical review,' *Environment and Planning A*, 23, pp. 705–30.

Groak, S. (1992). *The Idea of Building*, London: Spon.

Gruneberg, S.L. (1989). 'Economic criteria', in Osbourn, D. *Components*, 3rd edn, London: Mitchells.

Gruneberg, S.L. (ed.) (1996). *Responding to Latham*, Ascot: CIOB.

Gruneberg, S.L. (1997). *Construction Economics: an introduction*, London: Macmillan.

Gruneberg, S.L. and Weight, D. (1990). *Feasibility Studies in Construction*, London: Mitchells.

Harvey, D. (1982). *The Limits to Capital*, Oxford: Blackwell.

Harvey, D. (1989). *The Urban Experience*, Oxford: Blackwell.

Harvey A.C. and Jaeger, A. (1993). 'Detrending, stylised facts and the business cycle', *Journal of Applied Econometrics*, 8, pp. 231–41.

Hay, D.A. and Morris, D.J. (1991). *Industrial Economics and Organisation; Theory and Evidence*, (2nd edn), Oxford: Oxford University Press.

Hayek, F (1939). *Profits, Interest and Investment*, London: Routledge.

Heal, G. and Silberston, A. (1972). 'Alternative managerial objectives: an exploratory note', *Oxford Economic Papers*, 24, pp. 137–50.

Healey, P. (1992). 'An institutional model of the development process', *Journal of Property Research*, 9, pp. 33–44.

Healey, P. and Barrett, S. (1990). 'Structure and agency in land and property development processes: some ideas for research', *Urban Studies*, 27, pp. 89–104.

Hicks, J. (1973). *Capital and Time: A Neo-Austrian Theory*, Oxford: Clarendon.

Hillebrandt, P.M. (1984). *Analysis of the British Construction Industry*, London: Macmillan.

Hillebrandt, P.M. (1985). *Economic Theory and the Construction Industry*, London: Macmillan.

Hillebrandt, P.M. and Cannon, J. (eds) (1989). *The Management of Construction Firms*, London: Macmillan.

Hillebrandt, P.M. and Cannon, J. (1990). *The Modern Construction Firm*, London: Macmillan.

Hillebrandt, P.M., Cannon, J. and Lansley, P. (1995). *The Construction Company in and out of Recession*, Basingstoke: Macmillan.

Hodgson, G. (1988). *Economics and Institutions: A Manifesto for a Modern Institutional Economics*, Cambridge: Polity Press.

Inter Company Comparisons, (1977). *Business Ratios Business Ratio Report: Building and Civil Engineering (Major)*. London: Inter Company Comparisons.

Isard, W. (1942). 'A neglected cycle: the transport/building cycle', *Review of Economics and Statistics*, 24, pp. 149–58.

Ive, G. (1983). *Capacity and Response to Demand of the Housebuilding Industry*, London: UCL.

Ive, G. (1990). 'Structures and strategies: an approach to international comparison of industrial structures and corporate strategies in the construction industries', *Habitat International*, 14, pp. 45–58.

Ive, G. (1994). 'A theory of ownership types applied to the construction majors', *Construction Management and Economics*, 12, pp. 349–64.

Ive, G. (1995). 'Commercial architecture', in Borden, I. and Dunster, D. (eds) *Architecture and the Sites of History*, Oxford: Butterworth.

Ive, G. (1996). 'Innovation and the Latham Report', in Gruneberg, S. (ed.), *Responding to Latham*, Ascot: CIOB.

Ive, G. and Gruneberg, S. (2000). *The Economics of the Modern Construction Sector*, London: Macmillan.

Ive, G. and McGhie, W. (1983). 'The relation of construction to other industries and to the overall labour and accumulation process', *Production of the Built Environment, Proceedings of the Bartlett International Summer School*, London: UCL, 4, pp. 3-3 – 3-12.

Jackson, D. (1982). *Introduction to Economic Theory and Data*, London: Macmillan.

Janssen, J. (1983). 'The formal and real subsumption of labour to capital in the building process', *Production of the Built Environment, Proceedings of the Bartlett International Summer School*, London: UCL, 4, pp. 2-2 – 2-10.

Kalecki, M. (1971). *Selected Essays on the Dynamics of the Capitalist Economy*, Cambridge: Cambridge University Press.

Kay, N.M. (1984). *The Emergent Firm: Knowledge, Ignorance and Surprise in Economic Organisation*, London: Macmillan.

Keynes, J.M. (1936). *The General Theory of Employment, Interest and Money*, London: Macmillan.

Kindleberger, C. (1978). *Manias, Panics and Crashes: A History of Financial Crises*, New York: Basic Books.

Knight, F.H. (1933). *Risk, Uncertainty and Profit*, London: LSE.

Kotz, D.M., McDonough, T. and Reich, M. (eds) (1994). *Social Structures of Accumulation: The Political Economy of Growth and Crisis*, Cambridge: Cambridge University Press.

Kuznets, S. (1967). *Secular Movements in Production and Prices*, New York: A.M. Kelley.

Langlois, R.N. (1984). 'Internal organization in a dynamic context: some theoretical considerations', in Jussawalla, M. and Ebenfield, H. (eds), *Communication and Information Economics: New Perspectives*, Amsterdam: North-Holland.

Latham, M. (1994). *Constructing the Team*, London: HMSO.

Lee, F. (1999). *Post Keynesian Price Theory*, Cambridge: Cambridge University Press.

Leopold, E. and Bishop, D. (1983). 'Design philosophy and practice in speculative housebuilding: parts 1 and 2', *Construction Management and Economics*, 1, pp. 119–44, 233–68.

Lewis J.P. (1965). *Building Cycles and Britain's Growth*, London: Macmillan.

Linder, M. (1994). *Projecting Capitalism: A History of the Internationalisation of the Construction Industry*, London: Greenwood.

Lipietz, A. (1987). *Mirages and Miracles: The Crisis of Global Fordism*, London: Verso.

Littler, C. (1982). *The Development of the Labour Process in Capitalist Societies*, Aldershot: Gower.

Loasby, B.J. (1976). *Choice, Complexity and Ignorance: An Enquiry into Economic Theory and Practice of Decision Making*, Cambridge: Cambridge University Press.

Maddison, A. (1991). *Dynamic Forces in Capitalist Development*, Oxford: Oxford University Press.

Mandel, E. (1980). *Long Waves of Capitalist Development*, Cambridge: Cambridge University Press.

Marglin, S.A. (1974). 'What do bosses do? The origins and functions of hierarchy in capitalist production', *Review of Radical Political Economics*, 6, pp. 60–112.

Marglin, S.A. (1984). 'Knowledge and power', in Stephen, F.H. (ed.), *Firms, Organization and Labour*, London: Macmillan.

Marglin, S.A. and Schor, J.B. (1992). *The Golden Age of Capitalism*, Oxford: Clarendon Press.

Marx, K. (1970). *Capital: Volume 1* (trans. by Moore and Aveling from 3rd German edn; first published in English in 1887), London: Lawrence & Wishart.

Massey, D. and Catalano, A. (1978). *Capital and Land: Landownership by Capital in Great Britain*, London: Arnold.

Masterman, J.W.E. (1992). *An Introduction to Building Procurement Systems*, London: Spon.

Minsky, H. (1986). *Stabilising an Unstable Economy*, London: Yale University Press.

Morishima, M. (1984). *The Economics of Industrial Society*, Cambridge: Cambridge University Press.

Mueller, D.C. (1986a). *The Modern Corporation: Profits, Power, Growth and Performance*, Brighton: Wheatsheat.

Mueller, D.C. (1986b). *Profits in the Long Run*, Cambridge: Cambridge University Press.

Mueller, D.C. (1990). *The Dynamics of Company Profits: An International Comparison*, Cambridge: Cambridge University Press.

National Economic Development Council (1976). *Construction into the Early 1980s: The Implications for Manpower and Materials of Possible Levels and Patterns of Demand*, London: HMSO.

National Economic Development Council (1978). *'How flexible is construction?'*, London: HMSO.

Neale, A. and Haslam, C. (1994). *Economics in a Business Context*, 2nd edn, London: Chapman & Hall.

Needham, D. (1978). *Economics of Industrial Structure, Conduct and Performance*, Eastbourne: Holt, Rinehart & Winston.

Nelson, R. and Winter, S. (1982). *An Evolutionary Theory of Economic Change*, Cambridge, Mass: Belknap.

North, D.C. (1990). *Institutions, Institutional Change and Economic Performance*, Cambridge: Cambridge University Press.

Office for National Statistics (annually). *Business Monitor PA 1002, Production and Construction Inquiries, Summary Volume*, London: The Stationery Office.

Office for National Statistics (monthly). *Labour Market Trends*, London: The Stationery Office.

Office for National Statistics (annually). *United Kingdom National Accounts: the Blue Book*, London: The Stationery Office.

Office of National Statistics (Quarterly). *Economic Trends*, London: HMSO.

Okun, A.M. (1981). *Prices and Quantities – a macroeconomic analysis*, Oxford: Blackwell.

Park, W.R. (1979). *Construction Bidding for Profit*, New York: John Wiley.

Park, W.R. and Chapin, W.B. (1992). *Construction Bidding: Strategic Pricing for Profit*, Chichester: Wiley.

Penrose, E.T. (1959). *The Theory of the Growth of the Firm*, Oxford: Blackwell.

Penrose, E.T. (1980). *The Theory of the Growth of the Firm*, (2nd edn), Oxford: Blackwell.

Punwani, A. (1997). 'A study of the growth – investment – financing nexus of the major UK construction groups', *Construction Management and Economics*, 15, pp. 349–61.

Rawlinson, S. and Raftery, J. (1997). 'Price stability and the business cycle: UK construction bidding patterns', *Construction Management and Economics*, 15, pp. 5–18.

Reich, M. (1994). 'How social structures of accumulation decline and are built', in Kotz D.M. *et al.* (eds) *Social Structures of Accumulation*, Cambridge: CUP.

Reynolds, S. (1989). 'Kaleckian and post-Keynesian theories of pricing', in Arestis, R. and Kitromilides, Y. (eds), *Theory and Policy in Political Economy: Essays in Pricing, Distribution and Growth*, Aldershot: Edward Elgar.

Ricardo, D. (1973 edition). *On the Principles of Political Economy and Taxation*, London: Dent.

Richardson, G. B. (1960). *Information and Investment*, Oxford: Oxford University Press.

Robinson, J. (1933). *Economics of Imperfect Competition*, London: Macmillan.

Robinson, J. and Eatwell, J. (1973). *An Introduction to Modern Economics*, London: McGrawHill.

Rostow, W. W. (1971). *The Stages of Economic Growth*, Cambridge: Cambridge University Press.

Rowlinson, M. (1997). *Organisations and Institutions*, London: Macmillan.

Salter, W. (1969). *Productivity and Technical Change*, Cambridge: Cambridge University Press.

Sawyer, M. (1981). *Economics of Industries and Firms*, Beckenham: Croom Helm.

Schumpeter, J.A. (1934). *Theory of Economic Development*, Cambridge Mass., Harvard University Press.

Shackle, G.L.S. (1952). *Expectations in Economics*, Cambridge: Cambridge University Press.

Shackle, G.L.S. (1961). *Decision, Order and Time in Human Affairs*, Cambridge: CUP.

Shackle, G.L.S. (1967). *Years of High Theory*, Cambridge: Cambridge University Press.

Shackle, G.L.S. (1968). *Expectations, investment and income*, Oxford: Clarendon.

Shackle, G.L.S. (1972). *Epistemics and Economics*, Cambridge: Cambridge University Press.

Sherman, H. (1991). *The Business Cycle*, Princeton: Princeton University Press.

Simon, H.A. (1955). 'A behavioural model of rational choice', *Quarterly Journal of Economics*, 69, pp. 99–118.

Simon, H.A. (1957). *Models of Man*, New York: Wiley.

Simon, H.A. (1959). 'Theories of decision-making in economics and behavioural sciences', *American Economic Review*, 49, pp. 467–82.

Simon, H.A. (1976). *Administrative Behavior*, 3rd edn, New York: Macmillan.

Simon, H.A. (1979). 'Rational decision making in business organizations?' *American Economic Review*, 69, pp. 493–513.

Skinner, A.S. and Wilson, T. (1975). *Essays on Adam Smith*, Oxford: Oxford University Press.

Skitmore, M. (1989). *Contract Bidding in Construction*, Harlow: Longman.

Smith, A. (1976). *An Enquiry into the Nature and causes of the Wealth of Nations*, Campbell, R.H. and Skinner, A.S. (eds), Oxford: Clarendon.

Smith, N. (1984). *Uneven Development*, Oxford: Blackwell.

Smith, T. (1992). *Accounting for Growth*, London: Century Business.

Smyth, H. (1985). *Property Companies and the Construction Industry in Britain*, Cambridge: Cambridge University Press.

Sraffa, P. (1926). 'The laws of returns under competitive conditions', *Economic Journal*, 36, pp. 535–50.

Stumpf, I. (1995). *Competitive Pressure on Medium Sized Regional Contractors and their Strategic Responses*, MSc Dissertation, London: Bartlett School, UCL.

Sugden, J.D. and Wells, O. (1977). *Forecasting Construction Output from Orders*, London: University College Environmental Research Group.

Sweezy, P. (1939). 'Demand under conditions of oligopoly', *Journal of Political Economy*, 47, pp. 568–73.

Thomas, B. (1972). *Migration and Urban Development: A Reappraisal of British and American Long Cycles*, London: Methuen.

Thompson, G. (1986). *Economic Calculation and Policy Formation*, London: Routledge.

Thurow, L.C. (1976). *Generating Inequality*, London: Macmillan.

Thurow, L.C. (1980). 'Education and economic equality', in King, J.E. (ed.) *Readings in Labour Economics*, Oxford: OUP.

Turin, D. (1969). *The Construction Industry: its Economic Significance and its Role in Development*, London: UNIDO.

Turin, D. (1973). *Construction and development*, London: UCL.

Turin, D. (ed) (1975). *Aspects of the Economics of Construction*, London: Godwin.

Tylecote, A. (1992). *The Long Wave in the World Economy*, London: Routledge.

Van Duijn, J.J. (1983). *The Long Wave in Economic Life*, London: Allen & Unwin.

Varoufakis, Y. (1998). *Foundations of Economics*, London: Routledge.

Veblen, T. (1921). *The Engineers and the Price System*, New York: Harcourt Brace.

Veblen, T. (1964). *The Instinct of Workmanship*, New York: Augustus Kelly.

Vickrey, W. (1961). 'Counterspeculation, auctions, and competitive sealed tenders', *Journal of Finance*, 16, pp. 8–37.

Weisskopf, T.E. (1994). 'Alternative social structure of accumulation approaches to the analysis of capitalist booms and crises', in Kotz, D.M. *et al.* (eds) *Social Structures of Accumulation*, Cambridge: CUP.

Williamson, J. (1966). 'Profit, growth and sales maximisation', *Economica*, 33, pp. 253–6.

Williamson, O. (1975). *Markets and Hierarchies*, New York: Free Press.

Williamson, O. (1985). *The Economic Institutions of Capitalism*, New York: Free Press.

Wilson, T. and Andrews, P.W.S. (eds), (1951). *Oxford Studies in the Price Mechanism*, Oxford: Clarendon Press.

Winch, G.M. (1986). 'The labour process and labour markets in construction', *International Journal of Sociology and Social Policy*, 6, pp. 103–16.

Winch, G.M. (1989). 'The construction firm and the construction project: a transaction cost approach', *Construction Management and Economics*, 7, pp. 331–45.

Winch, G.M. (1995). 'Project management in construction: towards a transaction cost approach', *Le Groupe Bagnolet Working Paper*, 1, London: Bartlett School of Graduate Studies, London: UCL.

Winch, G.M. (1996a). 'The contracting system in British construction: the rigidities of flexibility', *Le Groupe Bagnolet Working Paper*, 6, London: Bartlett School of Graduate Studies, UCL.

Winch, G.M. (1996b). 'Contracting systems in the European construction industry', in Whitley, R. and Kristensen, P. (eds), *The Changing European Firm: Limits to Convergence*, London: Routledge.

Winch, G.M. (1998). 'The growth of self-employment in British construction', *Construction Management and Economics*, 16, pp. 531–42.

Winch, G.M. and Schneider, E. (1993). 'Managing the knowledge-based organisation: the case of architectural practice', *Journal of Management Studies*, 30, pp. 923–37.

Winch, G.M. and Campagnac, E. (1995). 'The organisation of building projects: an Anglo-French comparison', *Construction Management and Economics*, 13, pp. 3–14.

Wolfson, M.H. (1994). 'The financial system and the structure of accumulation', in Kotz, D.M. *et al.* (eds) *Social Structures of Accumulation*, Cambridge: Cambridge University Press.

Wood, A. (1975). *Theory of Profit*, Cambridge: Cambridge University Press.

Index